art of ENGINEERING

a handbook for problem solvers,
innovators & engineers

Art of Engineering

Copyright © 2024 Phil Underwood. All rights reserved.
First paperback edition printed 2024 in the United Kingdom

A catalogue record for this book is available from the British Library.

ISBN **978-1-910546-51-2**

No part of this book shall be reproduced or transmitted in any form or by any means, electronic or mechanical, including photocopying, recording, or by any information retrieval system without written permission of the publisher.

Published by artof

Although every precaution has been taken in the preparation of this book, the publisher and author assume no responsibility for errors or omissions. Neither is any liability assumed for damages resulting from the use of this information contained herein.

Version 1.02
January 2024

contents

intro — 5

part [i]

a HANDBOOK for problem solvers — 17
the mind of an engineer — 19
1. design — 29
2. build — 39
3. maintain — 49
engineering projects — 59

part [ii]

engineers IN ACTION — 69
what engineers do — 71
C craftsman — 81
D designer — 105
R research scientist — 133
M engineering manager — 169
how engineer develop — 201

appendix of book uses — 211

art of — 221
index — 223

intro

the author

If you'd told the 14-years-old me that I would write a book and be an engineer, I would have laughed, because I had always **struggled with words** and didn't know what I wanted to do for a job. I became a railway apprentice, I didn't know anything about engineering but had done well in the technical drawing class at school.

The apprenticeship taught me how to solve problems but more importantly, **how to learn things quickly,** I had to because I didn't know what I was doing most of the time. Luckily for me, although I didn't think so at the time, we had to write about our work and learning in a logbook, so to avoid writing words I tried to fill the pages with pictures. I soon realised that this approach really helped, I got good at fixing things and completed my apprenticeship in two years instead of four. I then got offered job opportunities.

I progressed from the shop floor to become a Designer, then a Research Scientist and Chartered Engineer. The path was not straight and was full of barriers and opportunities to grab (or be created). After a while in a job, I would get fed up and look for another challenge, I would always **record my achievements** and use these logbooks to help me get other jobs ! I knew when I started something new that I would feel out of my depth, but if I had a go and logged what I'd learnt I would get good at it (and the job).

Other people started to see that I could solve most problems and started to ask me if I help them solve their problems. Their **challenges were not engineering problems**, but I had a go and the engineering way of thinking worked and without knowing it I became a sort of troubleshooter. Then other people in other industries and fields started asking if I could help them, they said I helped them see things clearly, for the next 30+ years I would solve business problems and develop people. Then to help more people I started to write books.

the secrets of engineers

insight into a range of **real jobs**

see how **real problems** are solved

engineering is **a way of thinking** that can help you solve any problem

a handbook

The art of engineering is a pair of books in one, the Handbook for Problem Solvers and Innovators. And a Development Guide for Engineers.

Part [i] is a handbook that will help you solve problems, innovate and work on projects. It introduces you to a way of thinking and problem solving tools that engineers use. To illustrate how these skills and tools enable you to tackle genuine problems, links are provided for part [ii]

Part [ii] looks at engineers in action. It takes you inside the minds of engineers to see how real and complex problems are solved and how they develop into professional problem solvers. Sixty-six real-life problems show how the tools from the first chapters can be used to take on new challenges and help with continuous professional development.

> you don't have to read or use all the chapters,
> just move around & discover what you need

The art of engineering can also be used for learning and teaching purposes. An appendix of book uses shows links to curricula and professional competences, and how the book can support education, professional bodies and business innovation.

for the future

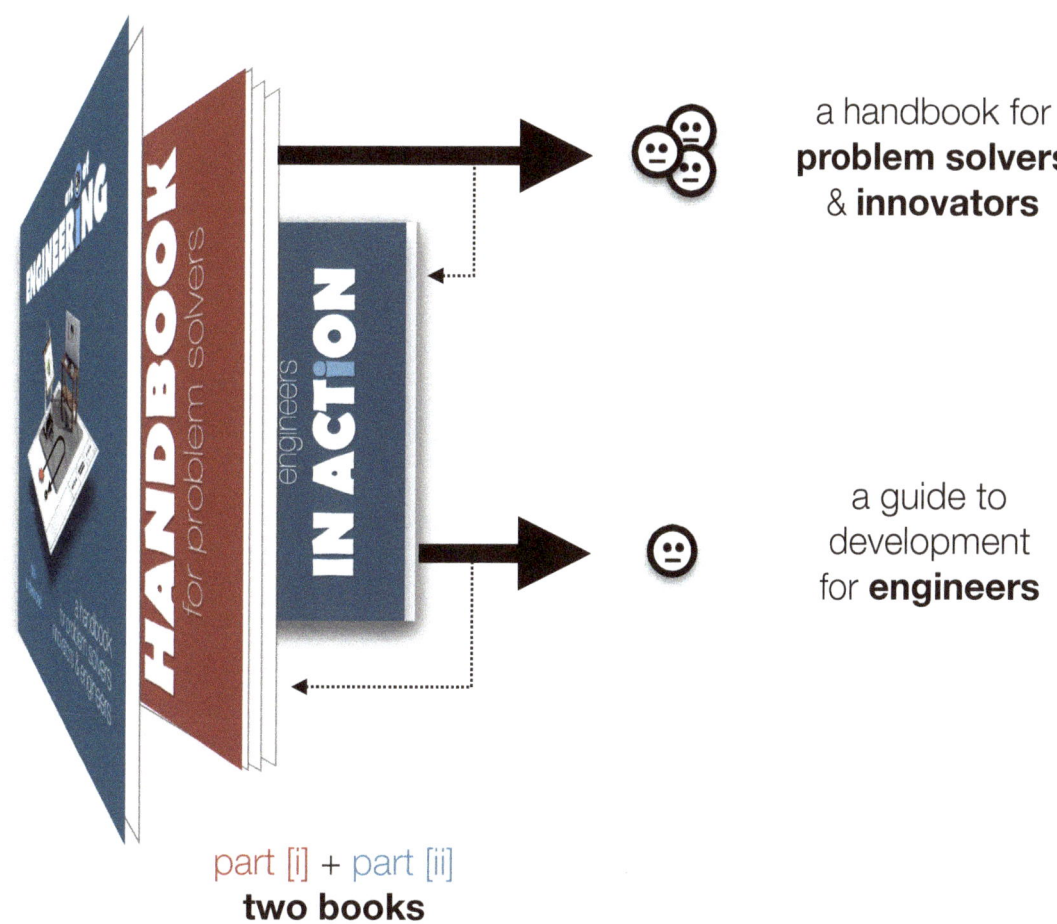

a handbook for **problem solvers** & **innovators**

a guide to development for **engineers**

part [i] + part [ii]
two books

the art of engineering

part [i]

The world is facing many challenges and there is a massive need to tackle today's problems and create a better tomorrow. The world desperately needs more problem solvers, innovators, and engineers.

Throughout history, engineers have been successful as the world's problem solvers, but how they do it and think is a bit of a mystery. Over the centuries engineers have developed proven methods and tools to bring ideas to life, and create solutions…. but what are they ?

helping you to **turn ideas into reality**

This book has two parts, that work together part [i] looks at 'what engineering is and how engineers approach problem-solving, part [ii] shows you 'what engineers do'. The two parts work together to help you turn ideas into reality and then see how engineers solve real problems.

In part (i) of the book we look at the problem-solving process, skills, and tools that engineers use have been honed over centuries. This knowledge is all brought together into a 'mind of an engineer' sheet that you can use to solve your problems.

Engineering is a way of thinking and an effective but different way of approaching problems. How an engineer thinks and the logical processes they employ uses the power of creativity (art) and reason (science).

a handbook for problem solving

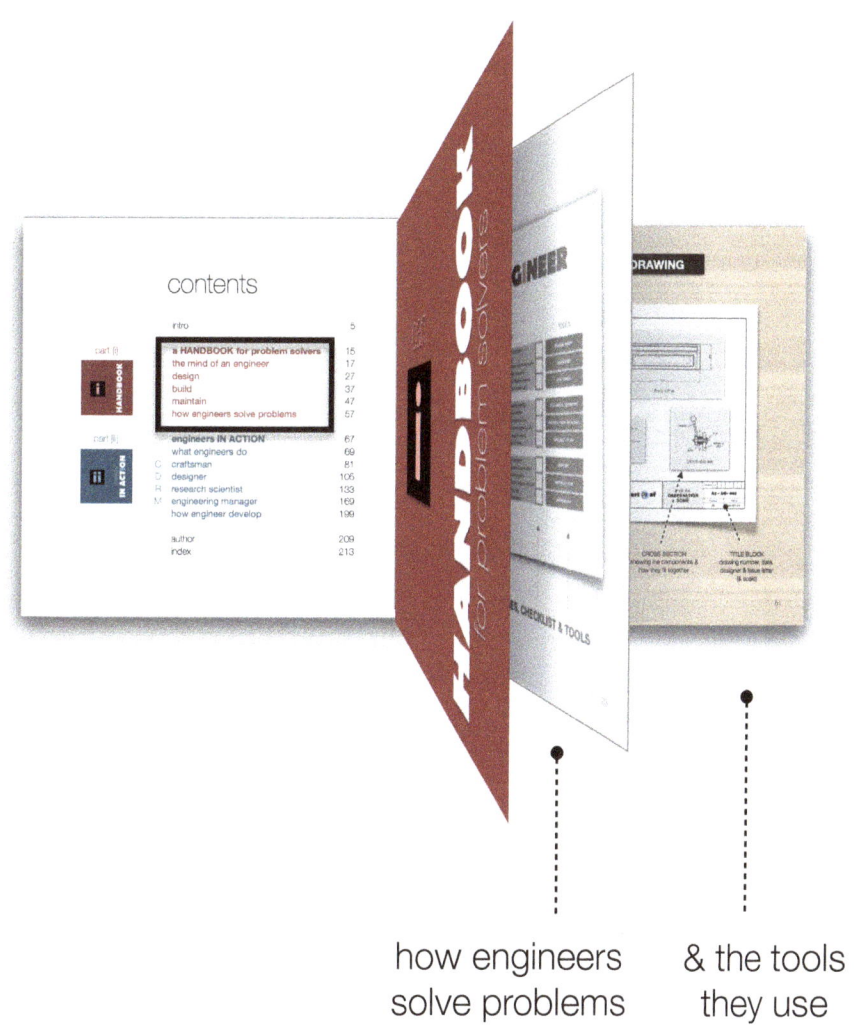

how engineers solve problems

& the tools they use

the art of engineering

part [ii]

In part (ii) we will look at some real engineering jobs, how the tools are used, and how they develop as problem solvers. Real problems and projects will be used to help you understand what happens when they are in full flow doing engineering.

To show the variety and levels of the work that engineers do we will focus on four different types of jobs, craftsman, designer, research scientist, and engineering manager. You'll be able to see their world and how they deal with the real-world challenges.

insights from real problem solving
and how to develop as a problem solver

Different engineering challenges will use a different mix of tools to help solve specific types of problems. I'm going to demonstrate how engineers deal with their problems, apply their judgement and the journey that ideas take in order to solve them

To give an insight into how engineers approach specific problems, examples of real work (polaroid pages) show the process from problem to solution and how some of the tools from part [i] are used.

Different problems need different problem solvers. We will look at how different four jobs at different levels and in different departments use the knowledge and tools from part (i). Different engineering jobs will have a specific mix of design, build, and maintain skills.

see engineers **in action**

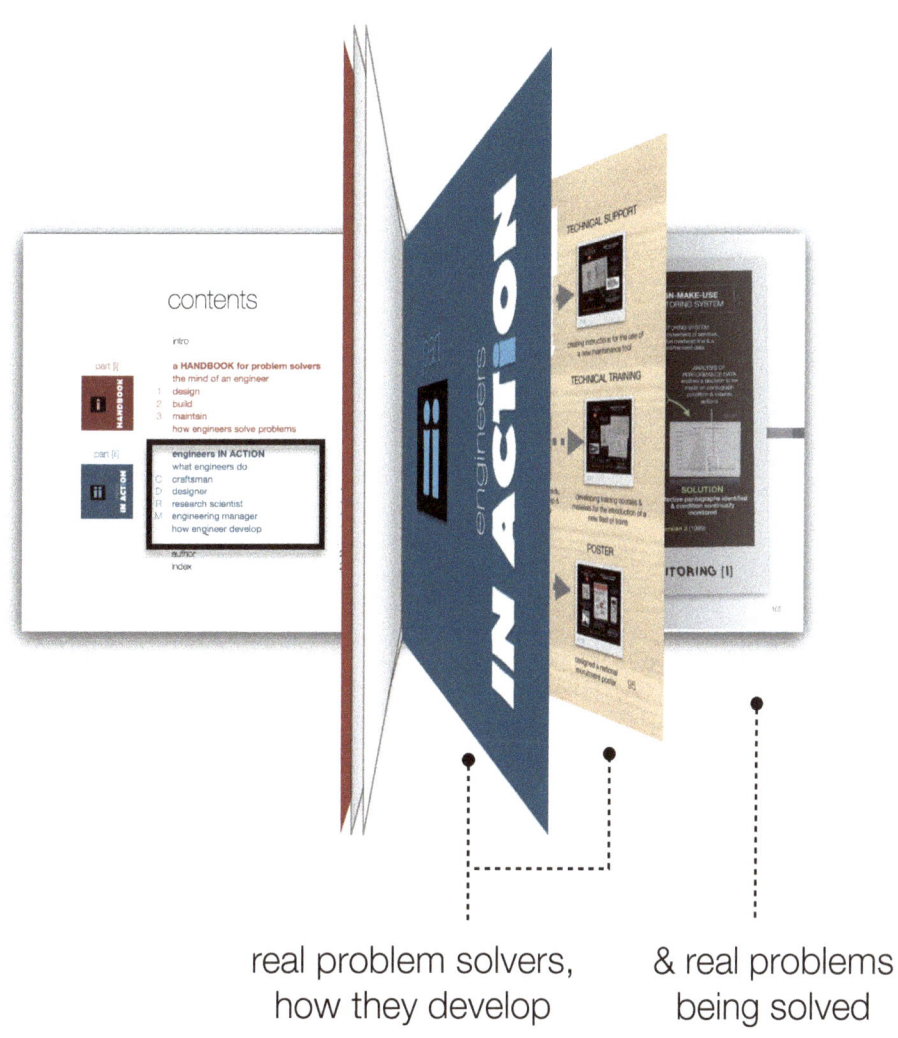

real problem solvers, how they develop

& real problems being solved

To help you discover the secrets of the world's problem solvers, the way they think, and the tools they use, we'll use the opposite picture. We will go inside the mind of an engineer and see how real problems are solved, and see what engineering looks like when they are in full flow doing engineering.

The arrows show how the thought or problem flows to a solution. Part [i] provides a "checklist" we can use to ensure that we have done all of the "required" steps, then nine tools to help answer nine key questions. Through a number of real problems and solutions, the tools are brought to life in Part [i].

equipping you
to innovate, do something new, or Improve an existing thing

Engineers approach and solve problems differently, they use drawing as a language, and use them to think aloud. They use drawing to think and communicate ideas, then use these drawings to navigate the countless problems on the idea-to-reality journey.

Engineers use the power of graphics in three basic ways: to visualise the idea, to communicate the idea so others can evaluate it, and to document the solution so the solution can be reliably reproduced and maintained.

Adopting an engineering mindset can help you in any field. We can borrow strategies from engineering to find inspired solutions to other challenges. A different way of thinking, not limited to Engineering...

a different approach

the art of engineering

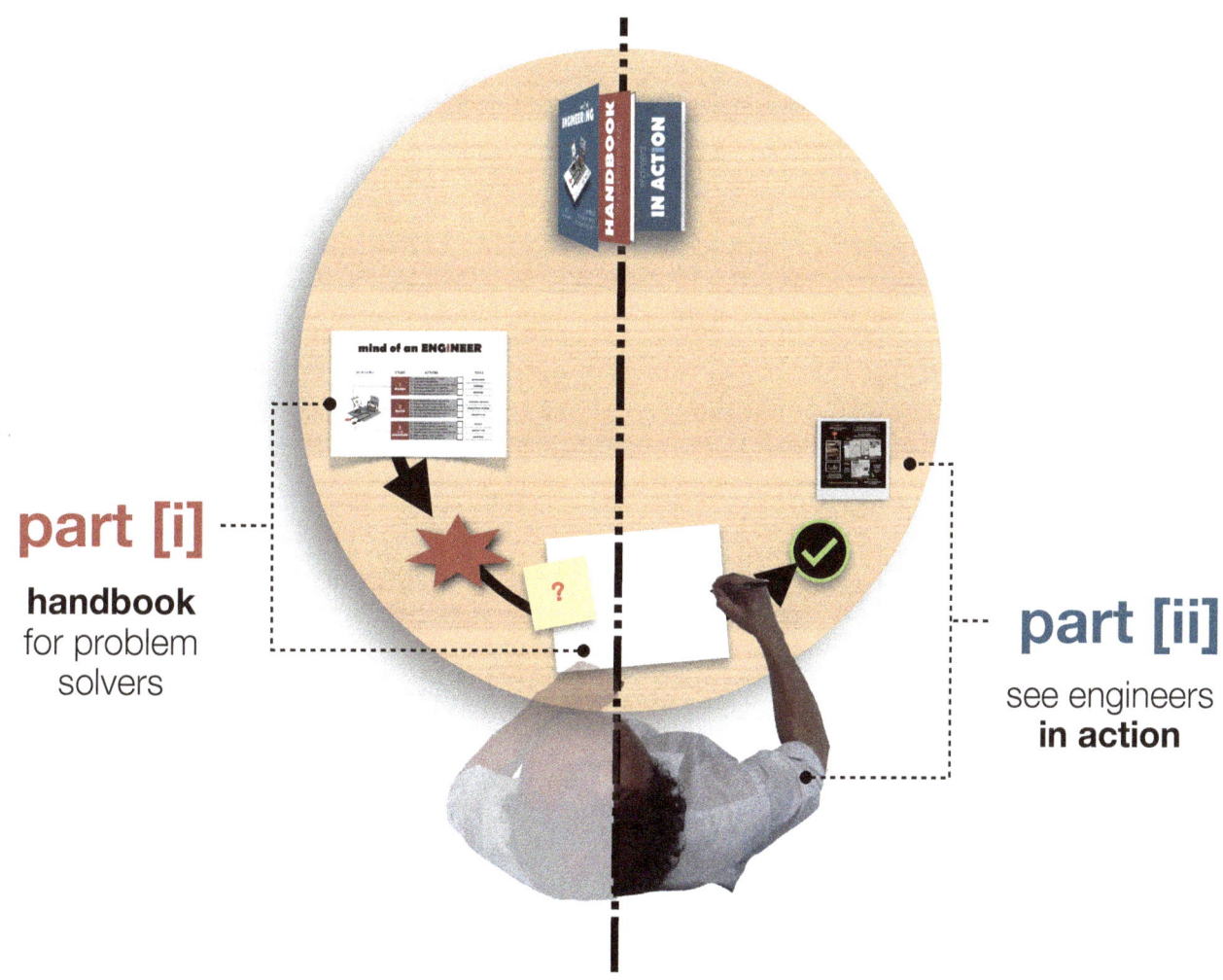

part [i]
handbook
for problem solvers

part [ii]
see engineers **in action**

I'm going to use some personal notebooks and drawings throughout the book to bring engineering practice to life. The old-school, pen and paper drawings help show that engineering is a creative activity and explain my thought process. But they also became a continuous record of my personal and professional development.

art of &
the author

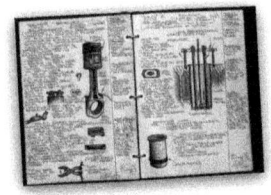 PERSONAL
NOTEBOOKS

Although the problems and projects shown here are from the rail industry and were done several years ago the principles and processes are still key to all engineering work today. They also illustrate the breadth and depth of professional engineering, the potential career paths and and show how exciting a career in engineering can be.

part

HANDBOOK for problem solvers

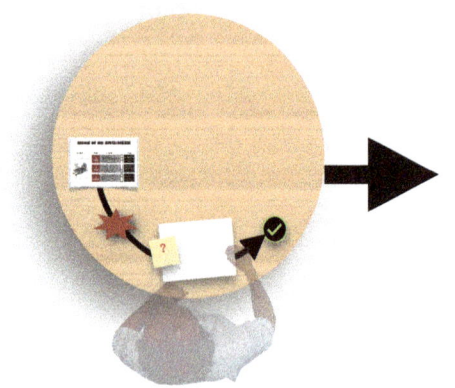

HANDBOOK FOR PROBLEM SOLVERS

You'll learn how to think and how to turn your ideas into reality in this part of the book. The five chapters are linked together to equip you with :

- ❏ A CHECKLIST for problem-solving
- ❏ Practical TOOLS to create solutions
- ❏ Help plan & manage a PROJECT

HELPING YOU :

- ❏ develop & communicate an idea
- ❏ run a project
- ❏ present a proposal

The chapters help you create output documents for education (pages 210 & 212), pitching business projects , and helping with professional development (page 214).

the mind of an engineer

the worlds problem solvers

Throughout history, engineers have been successfully solving difficult problems and creating the future for 200+ years. You may not have noticed we live in a world shaped by engineers and that engineering is part of life for us all. Their solutions are all around us and make modern life possible.

Engineering is behind everything we use in the home, at the office and to help us move around, transport. Engineering is everywhere and we often take it for granted.

making modern life possible &
improving the world

The world has always needed engineers to move the world forward such as Watt, Stephenson, Brunel, Edison, Whittle etc They have continually turned ideas into reality and have invented the future.

The focus for engineers is on improving an existing solution or innovation to create something fundamentally new. Taking a revolutionary approach to a problem or a small, incremental step to make something better.

Let's zoom in and see what they do…

improving the world

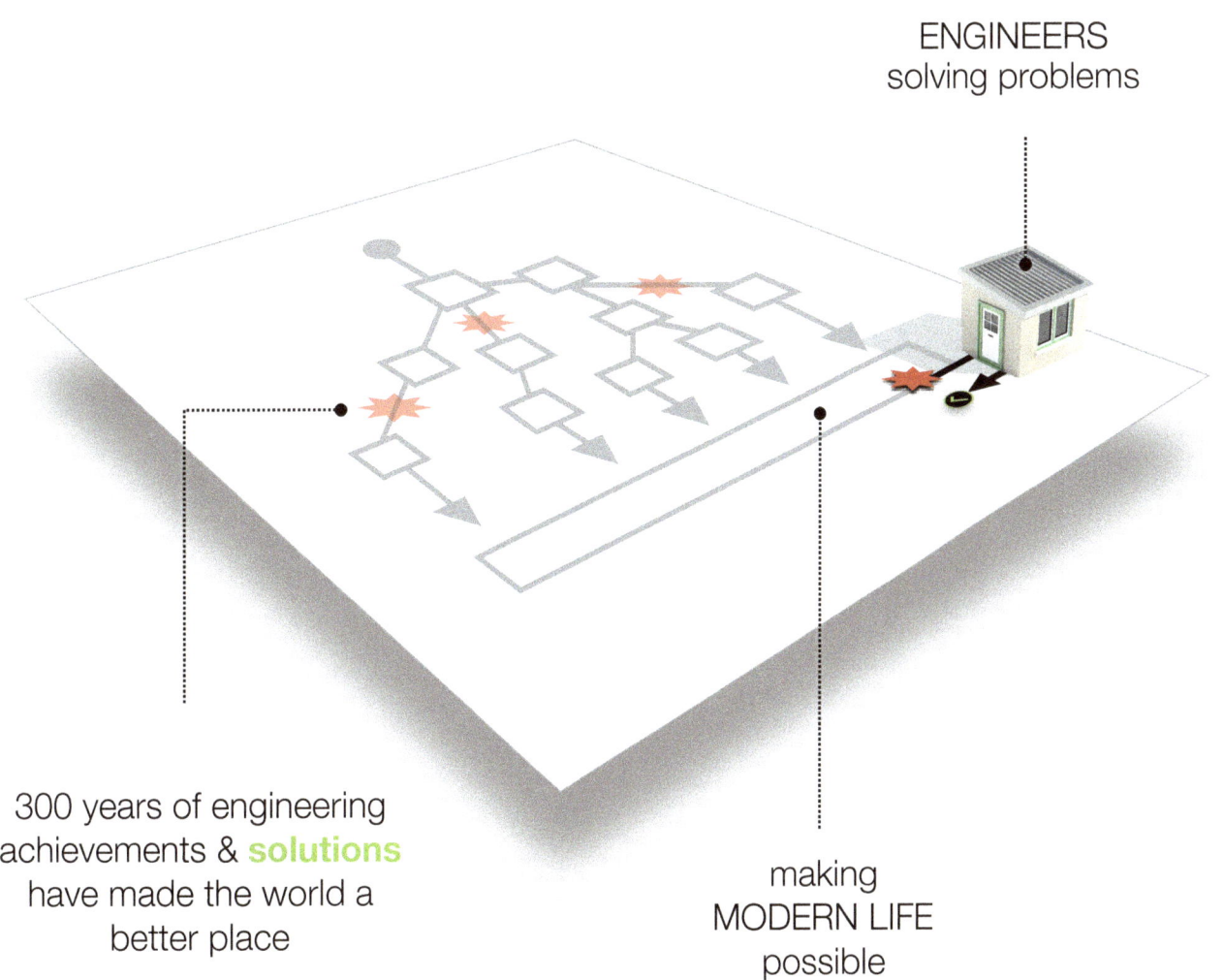

ENGINEERS
solving problems

300 years of engineering achievements & **solutions** have made the world a better place

making
MODERN LIFE
possible

what is engineering

Engineers are the world's problem solvers, but how they do it is a bit of a mystery, we don't know enough about what engineering is, how they think, and what they do.

The major functions of engineering are designing, building, and maintaining. They use the drawing board to design things and the workshop to build and maintain solutions.

transforming problems into opportunities & **turning ideas into reality**

Engineers see new problems as opportunities to do something better and improve the world. They improve and innovate, and keep existing solutions in existence. They repair and maintain things, they detect faults and prevent failures and prolong the lives of things.

Although engineering is a highly scientific and technical profession, creativity is embedded in every step of the process used. No two problems are the same and for this reason, creativity is needed to overcome problems.

Let's understand a bit more about the process they have developed…

an approach to problem solving

the DRAWING BOARD

the WORKSHOP

what engineers do

An engineer is a person who designs, builds, or maintains things. The way they do this is not always visible, so let's explore the set of activities that engineers go through.

The idea to reality, the end-to-end stages is a process proven and has been honed over the centuries. The design/build/maintain process here helps us see what engineers are doing :

[1] **design**
understanding the problem, then explain how the solution works (function) & what it looks like (form)

[2] **build**
demonstrate that there is a reliable & cost-effective way to manufacture the solution

[3] use & **maintain**
then prove that the ideas are useful, that they can be maintained & then tell the world about them

Engineering is about making real things that work and serve a purpose, they make stuff or make things work better. Although engineering problems vary in scope and complexity, the same general approach is applicable.

Let's now understand how the mind of an engineer works and the practical steps that they take..

designing, building & maintaining

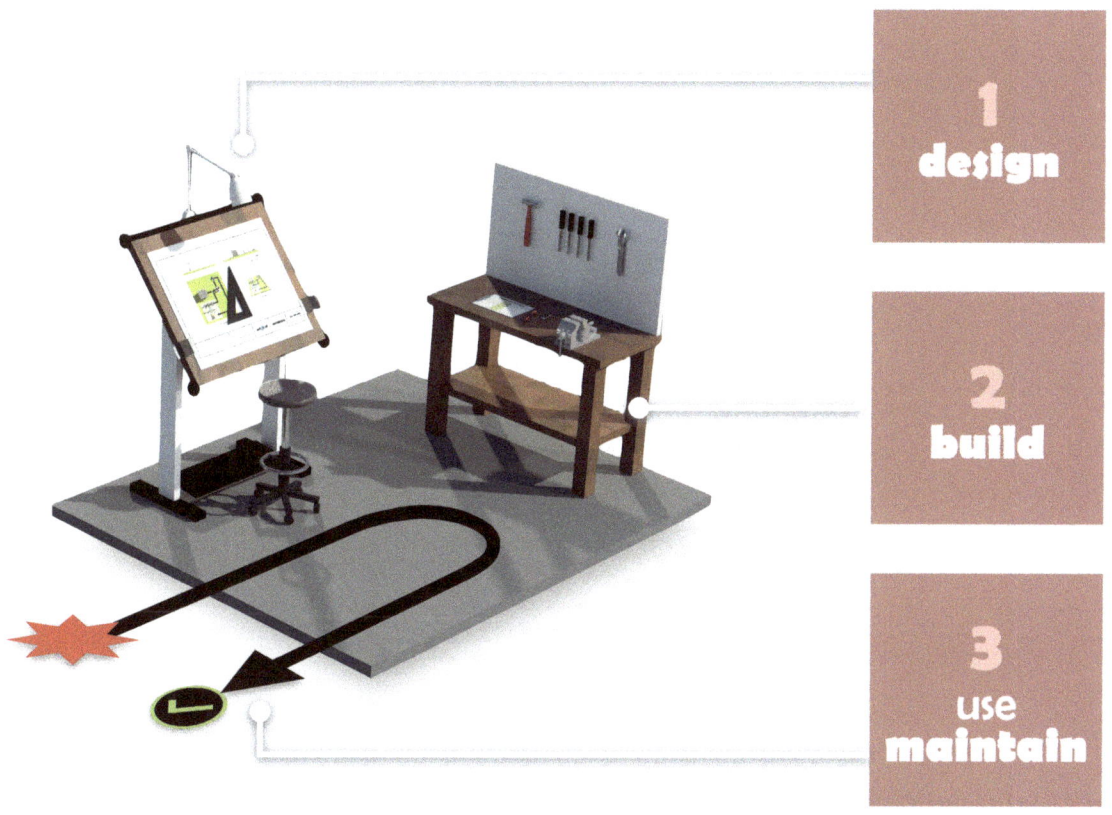

three STAGES

1 design

2 build

3 use maintain

how engineers solve problems

Engineering isn't a job, it's a way of thinking, a way of approaching problems, challenges, and opportunities. It's a logical and systematic process that enables the problem solver to navigate the countless unique problems on the idea-to-reality journey.

The 'mind of the engineer' model opposite describes the methodical steps that engineers use in creating functional products or processes, from the design brief to the finished work. It is a checklist for solving a problem or working on a project.

the mind of an engineer
a process for problem-solving

Each step can be seen as a task or skill that you need and can help you develop your problem solving abilities. The checklist is designed to help you plan and track your project's progress, as well as provide information on the skills that might be needed.

Over the centuries engineers have developed proven tools to help them solve problems quickly and effectively. There are nine tools in the note section that will help you turn your ideas into reality. The tools, together with real examples will be explained in detail and demonstrated throughout the book.

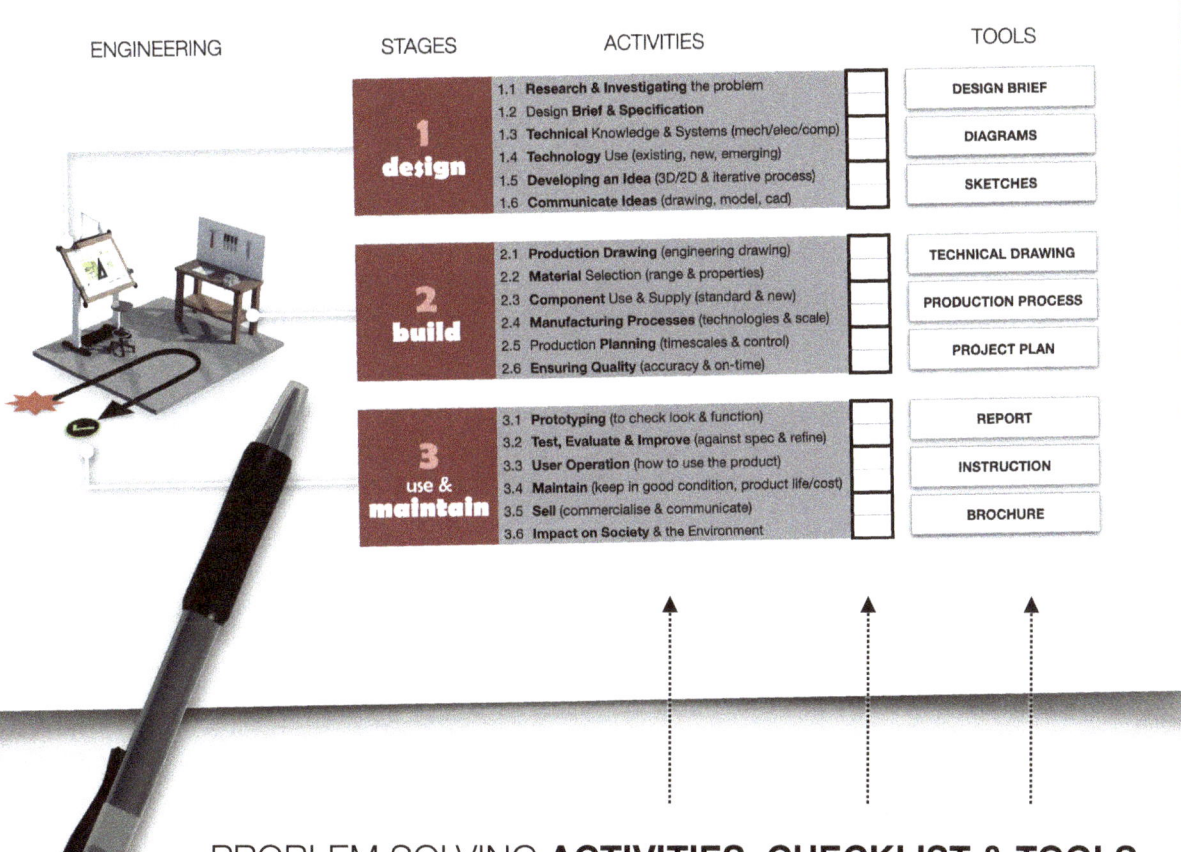

The modern world is confronting some new and formidable problems and needs novel solutions. We need people with the ability to create and innovate in a rapidly changing world.

The great thing about the way engineers solve problems is that it can be applied by anyone in their own lives and careers.

become a better problem solver

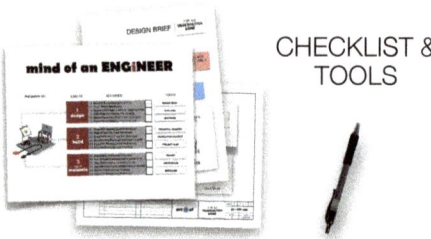

CHECKLIST & TOOLS

We can borrow strategies from engineering to find inspired solutions to our most pressing challenges.
Anyone can adopt an engineering way of thinking and become more successful

In the next three chapters, we will look closer at how engineers use nine questions and nine tools to break down a problem and find solutions.

1
design
STAGE

design skills & tools

The word "design" means to draw something, hese sketches show how something works and what it looks like. But before we begin to draw, we have to know what problem we're dealing with.

Engineers must know that three key questions need to be answered when designing their solutions :

> understanding the problem ?
> how does it work ?
> what will it look like ?

Understanding **the problem** is an important first step in a design process. By developing a deep understanding of how and why things work, we can then define the requirements that a solution need to meet.

These requirements help to define the functioning of a solution, which is looking at **how it will work** and what technologies are being employed. The initial concept is developed by the designers using scientific principles, technical knowledge and imagination.

Drawings are then used to develop the idea further and communicate **what it looks like**. Illustrations help give a realistic description of the finished product and hopefully help excite others.

the design stage

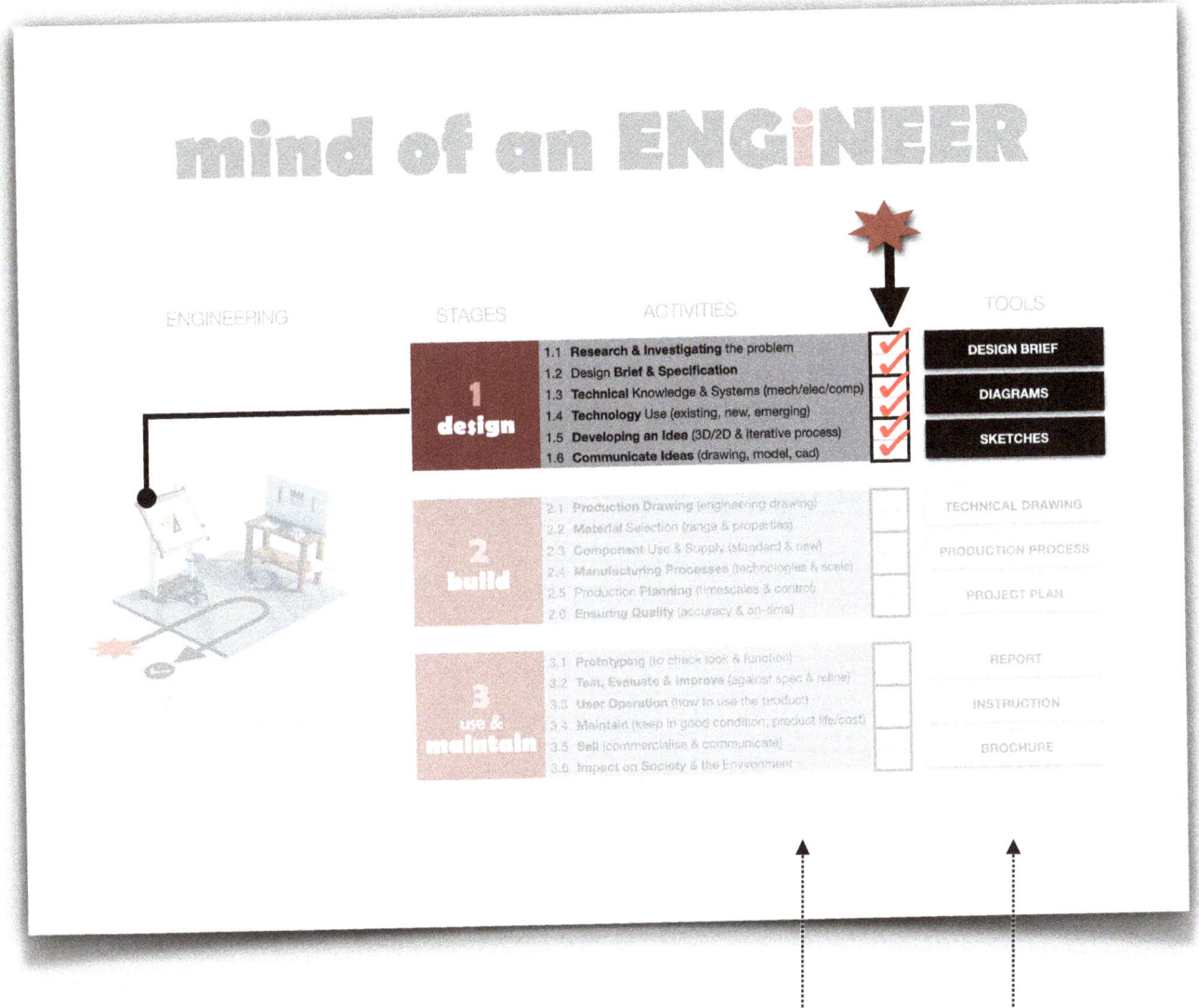

DESIGN ACTIVITIES & TOOLS

understanding the problem

... how to use a document to answer the QUESTION

WHAT IS THE PROBLEM ?

The first thing we do is to research and investigate the problem, breaking it into bits to specify what needs to be improved or created, and then describe the problem in a **design brief**.

1.1 Research & Investigating the problem

Designers carry out research and investigation to find out more about the design problems (or system), and the needs of the market, client, and/or end user.

1.2 Design Brief & Specification

A design brief is a short description of a design problem and how it is to be solved. A design specification can also be written to list any measurable design criteria that a product or system must meet, this will help when it comes to evaluating any solutions.

DESIGN BRIEF

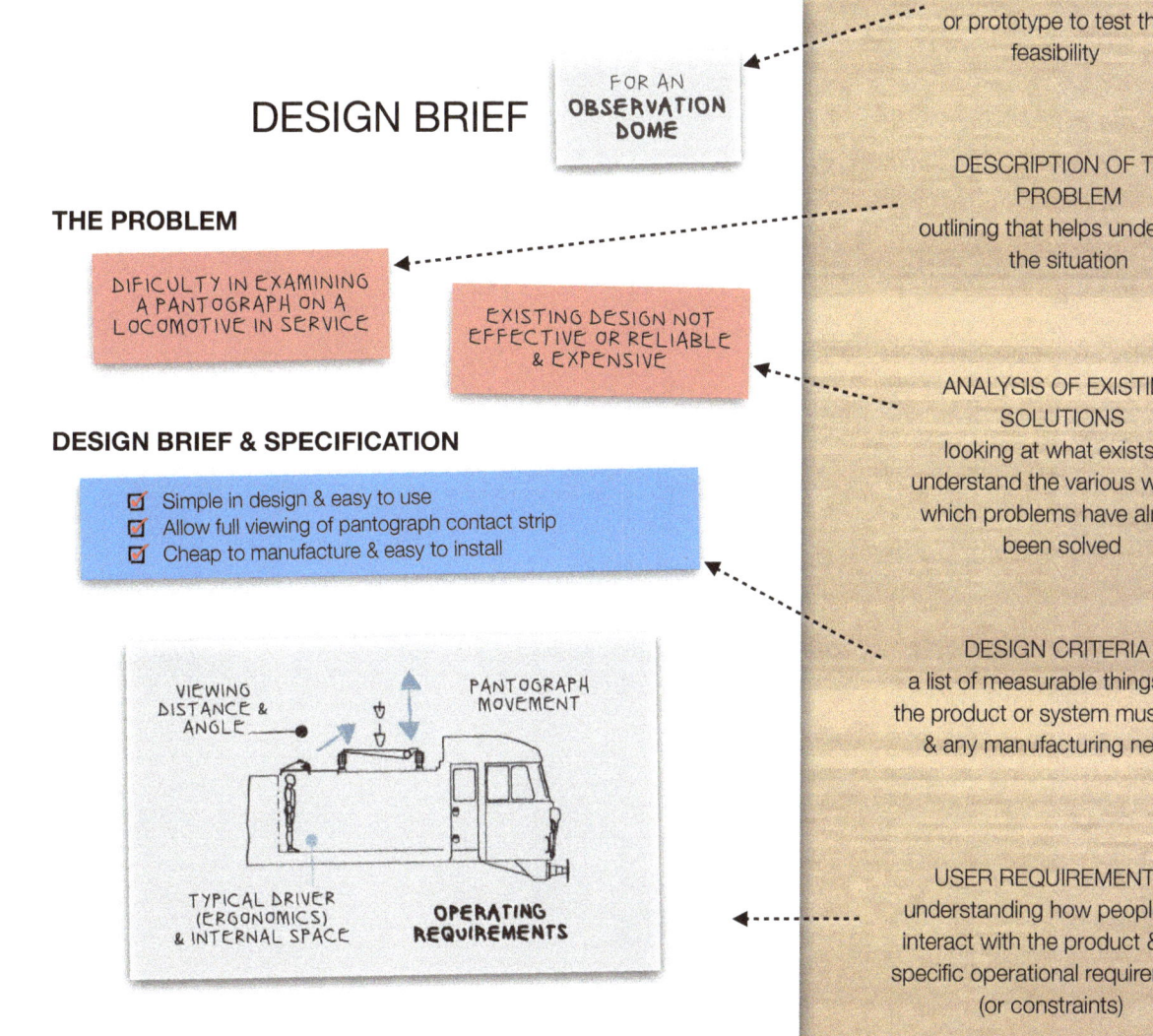

information from a REAL PRODUCT used to show the tool being used, see R13, R14 & R15

how does it work

... how to use pictures to answer the QUESTION

When engineers look at things they see systems and technologies. They then use **diagrams** to create something better and then apply their understanding of science to predict its performance.

1.3 **Technical** Knowledge & Systems (mech/elec/comp)

Most products and systems involve a mixture of mechanical, electrical, and computer systems, and an energy source. To show how a system will work we can use schematic diagrams and represent elements of a system using standard symbols rather than pictures.

1.4 **Technology** Use (existing, new, emerging)

Innovation often requires making good use of existing, new, or emerging technology. However new or improved materials or manufacturing processes can also drive innovation.

DIAGRAM

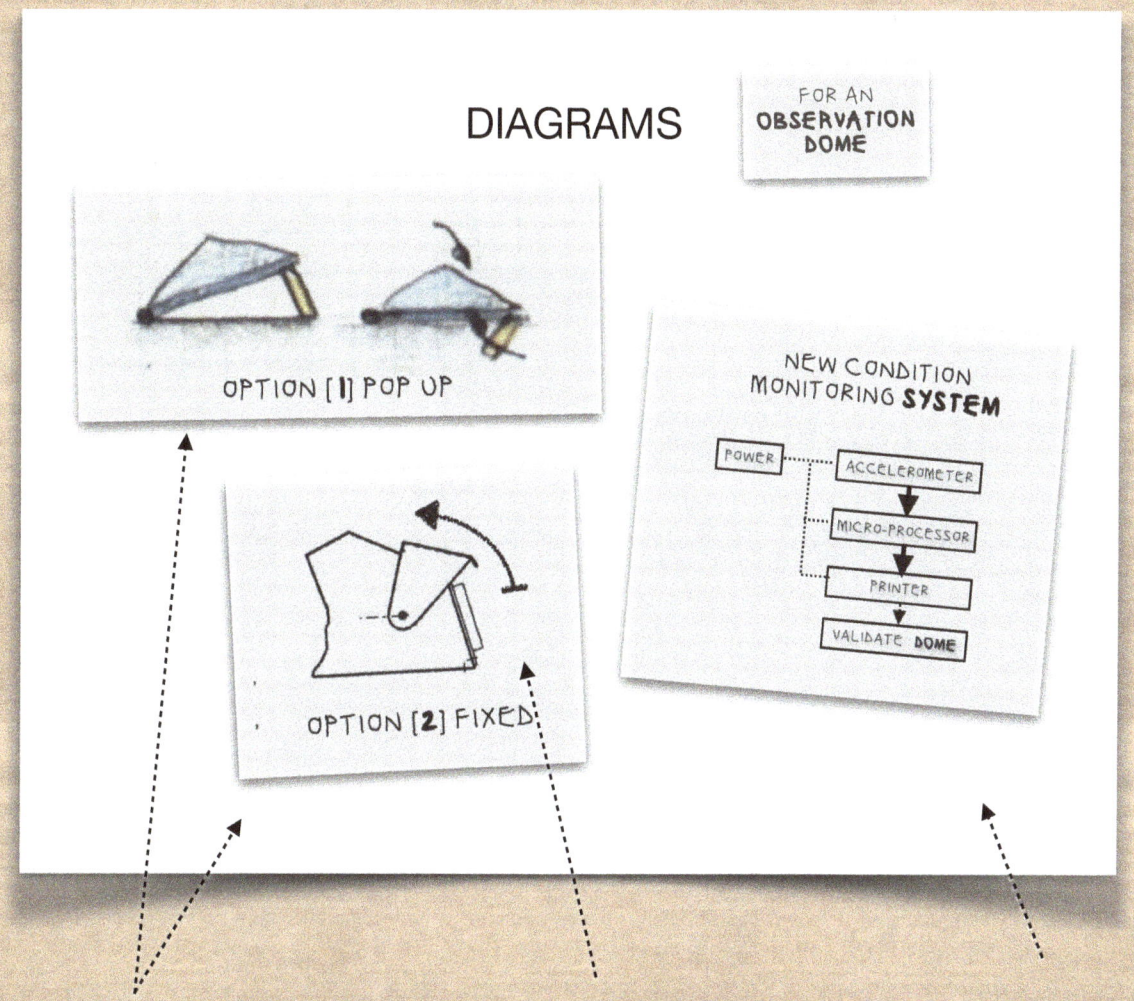

GENERATE CREATIVE IDEAS
understanding user needs, to help develop some initial concepts, that can then be developed further

TECHNOLOGY USED
understanding the use of materials, structural elements & how mechanical systems enable movements

SYSTEM SCHEMATIC DIAGRAM
this diagram shows how a new condition monitoring system will work & the additional requirement for visual observation

what does it look like

... how to use a picture to answer the QUESTION

Engineers use **sketches** to help them see their thinking and check that the idea will work and that it can be built. The drawings are then used to communicate the idea to others and create interest or generate other ideas.

| 1.5 | **Developing an Idea** (3D/2D & iterative process) |

Freehand sketching is a good way to get initial thoughts and ideas down on paper. Designers also make models of their ideas to check how they will look and function in 3D.

| 1.6 | **Communicate Ideas** (drawing, model, CAD) |

In order to be able to provide sufficient information about the design, drawings can be marked and labelled. These initial drawings give you with a quick method of communicating and gaining feedback on ideas. By testing ideas and getting feedback from others ensures that problems and improvements are found early in the process.

SKETCH

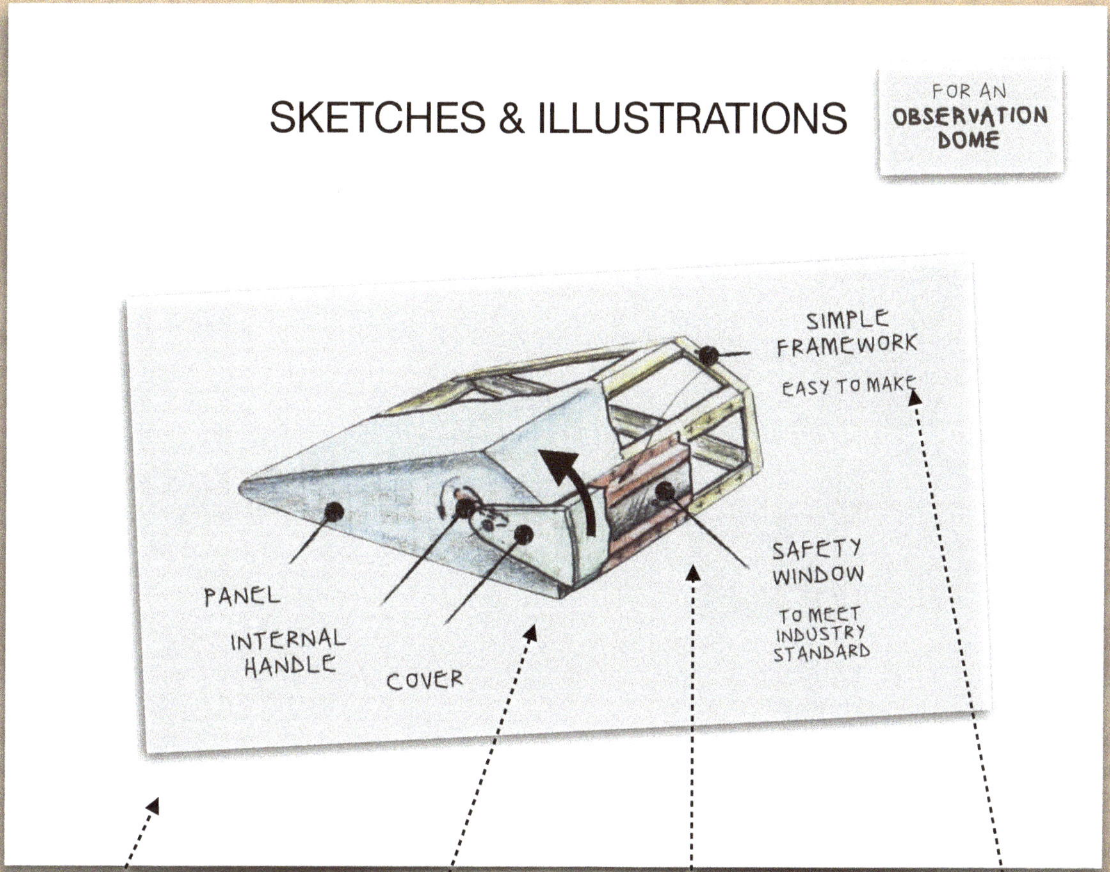

SKETCHES & ILLUSTRATIONS

FOR AN OBSERVATION DOME

- PANEL
- INTERNAL HANDLE
- COVER
- SIMPLE FRAMEWORK — EASY TO MAKE
- SAFETY WINDOW — TO MEET INDUSTRY STANDARD

3D DRAWING
to describe the design of different parts & assembly

OPERATION
making it easy to use, how it works & the moving parts

SAFETY
the user is protected from danger or injury

MANUFACTURING CONSIDERATION
cost/time produce & simple production process

design tools in action

DESIGN BRIEFS
- [ ] New System **R8, R12, R13**
- [ ] Maintenance Process **M4, M5**
- [ ] Maintenance Tool **R19**
- [ ] IT System **M6**

DIAGRAMS
- [ ] Fault Finding Schematic **C8, C9, R6**
- [ ] Technical Training **C11**
- [ ] New System Block Diagram **R7**
- [ ] Modelling a System **R17, M2**

SKETCHES
- [ ] 3D CAD Drawing **R11**
- [ ] External View **D12**
- [ ] Internal View **D10**
- [ ] Vision Sketches **R4, R16, R18**

Now you understand how to design something and three tools that can help you to develop an idea. Practice using the design tools [1], [2], and [3] will help you develop your design skills.

2
build
STAGE

build
skills & tools

The word build means to make something by putting parts and materials together. Before we go ahead and try to make something we need to be able to demonstrate that there is a reliable means of manufacture.

Three key questions are answered by engineers in order to develop their successful solutions :

information required by the makers ?
how will it be made ?
how long will it take ?

The designer has to translate ideas into **the technical information required** by a workshop. Technical drawings are used because they can store a large number of decisions and allow precise communication.

The steps through which raw materials are transformed into the final product is **the manufacturing process**. Manufacturing helps us make something on a large scale by using machinery.

Production planning is used to help allocate resources and organise the activities of people, materials, and equipment. It is planning that helps things flow and be carried out efficiently.

the build stage

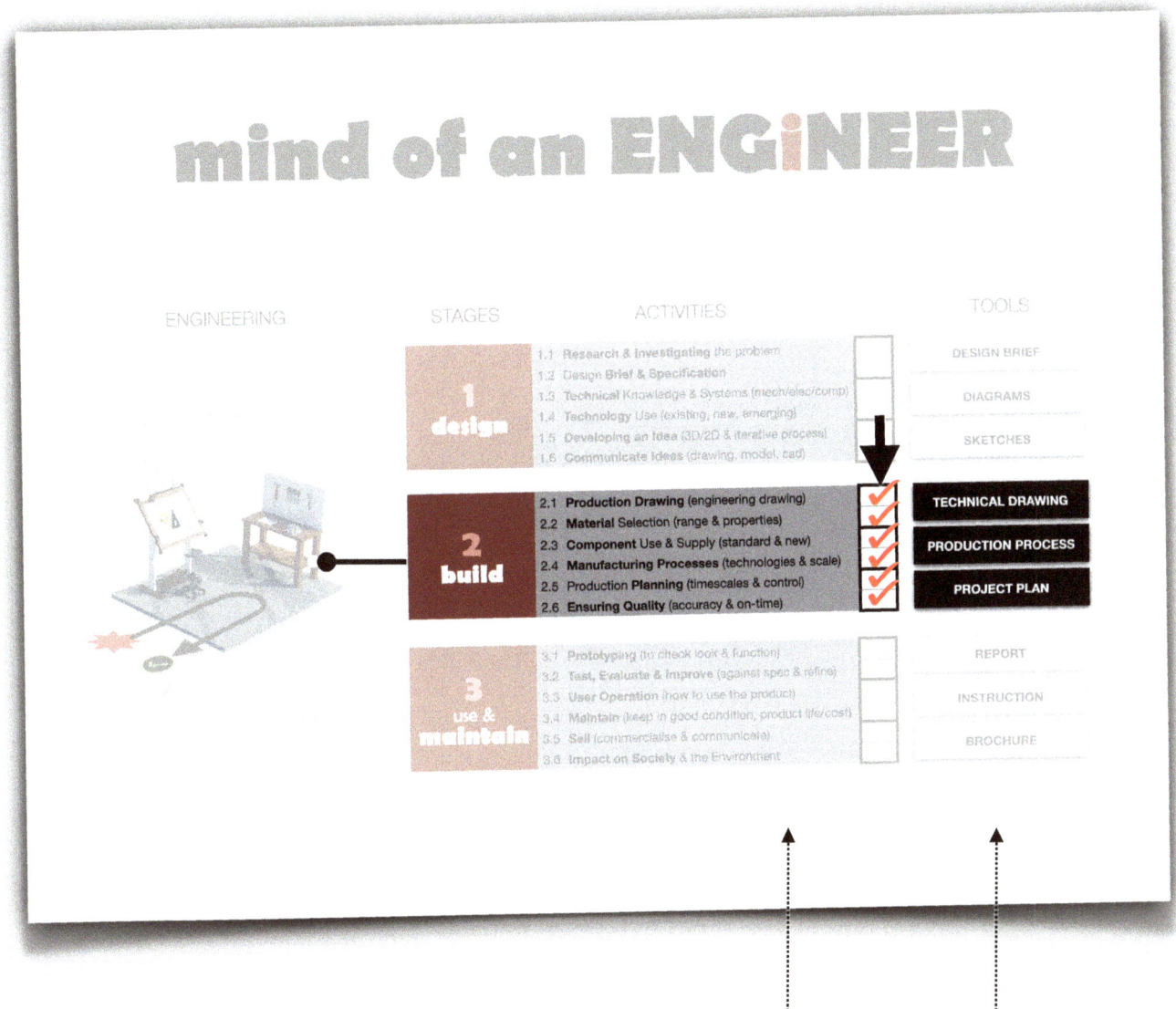

BUILD **ACTIVITIES** & **TOOLS**

information required by the makers

... how to use pictures to answer the QUESTION

We need to convey information about our idea or object to the maker. We do this by producing a scaled **technical drawing** that allows it to be made by anyone and anywhere in the world.

2.1 Production Drawing (engineering drawing)

Technical drawings communicate very specific and detailed information about size, material, assembly, and connections. They use different views and projections to communicate a 3D picture of the product in 2D. The drawings give a continuous picture of a product as it develops and have the capacity to store lots of decisions.

2.2 Material Selection (range & properties)

The main goal of material selection is to minimise cost while meeting product performance goals. Finding the best material for a given application begins with the required properties. Different materials ie metals, wood, polymers, etc will have quite different properties and need different manufacturing processes.

TECHNICAL DRAWING

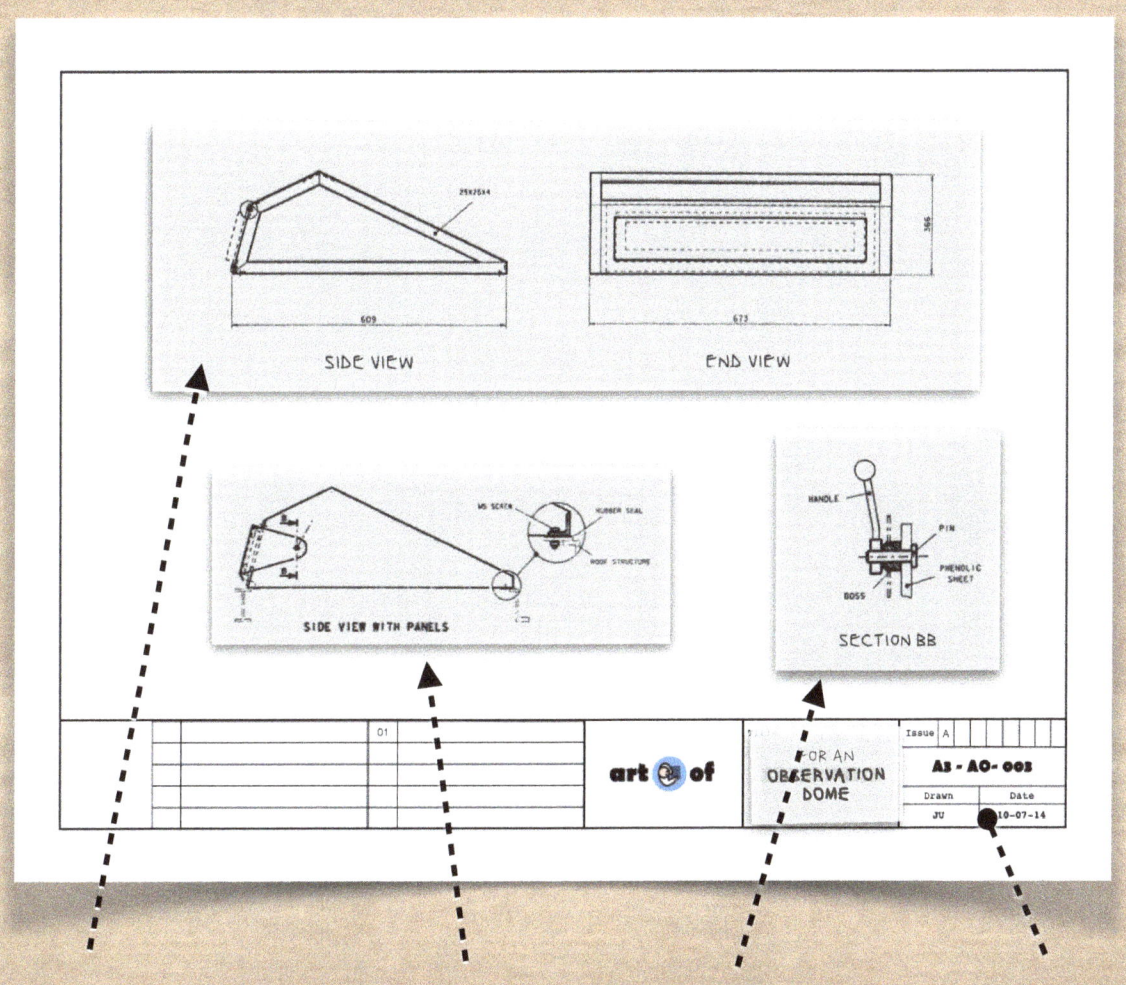

PROJECTION
scale drawing of side & end views, with dimensions & material details

INSTALLATION
a drawing to show how the dome fits onto the roof of the locomotive

CROSS SECTION
showing the components & how they fit together

TITLE BLOCK
drawing number, date, designer & issue letter (& scale)

how will it be made

... how to use stick-its & logic to answer the QUESTION

Things can often be made in a variety of different ways. The **production process** describes the steps through which materials are turned into products.

HOW WILL IT BE MADE ?

| 2.3 | **Component** Use & Supply (standard & new) |

Designers normally try to use standard components as much as possible because they are cheaper and readily available. The parts are normally available in a range of standard sizes and can be used in a variety of different products.

| 2.4 | **Manufacturing Processes** (technologies & scale) |

The steps through which raw material are transformed into a final product is the manufacturing process. Different materials will require the appropriate tools and equipment to carry out the processes. But the number of identical products to be made and the scale of manufacturing will also help us identify the manufacturing equipment we might use.

PRODUCTION PROCESS

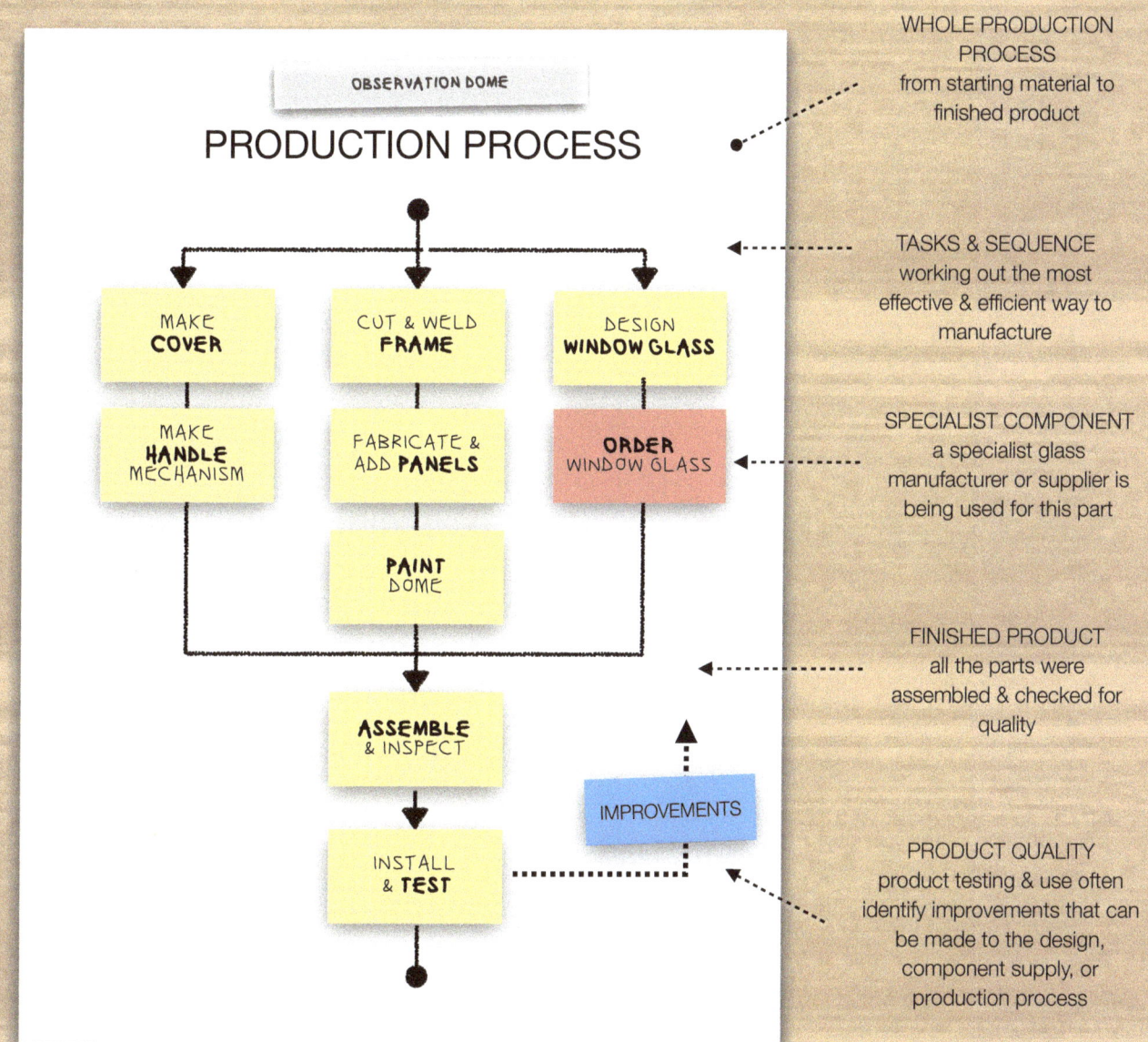

time & resources to make it

... how to use stick-its & logic to answer the QUESTION

To understand how long it will take to produce the product we need to plan all the required activities. A **project plan** lets us timeline what needs to be done and track progress.

| 2.5 | Production **Planning** (timescales & control) |

It takes a lot of resources and time to make something. To predict how long it might take we need to list all the tasks that need to be done, then sequence them so and set milestones. This becomes a project plan, it lets us find the best path, monitor progress, and control the project

| 2.6 | **Ensuring Quality** (accuracy & on-time) |

Accuracy is extremely important when manufacturing things, just a small deviation from the specified measurement can result in a poor product. Measurement and production aids such as jigs, templates, and patterns help to achieve accuracy and repeatability.

PROJECT PLAN

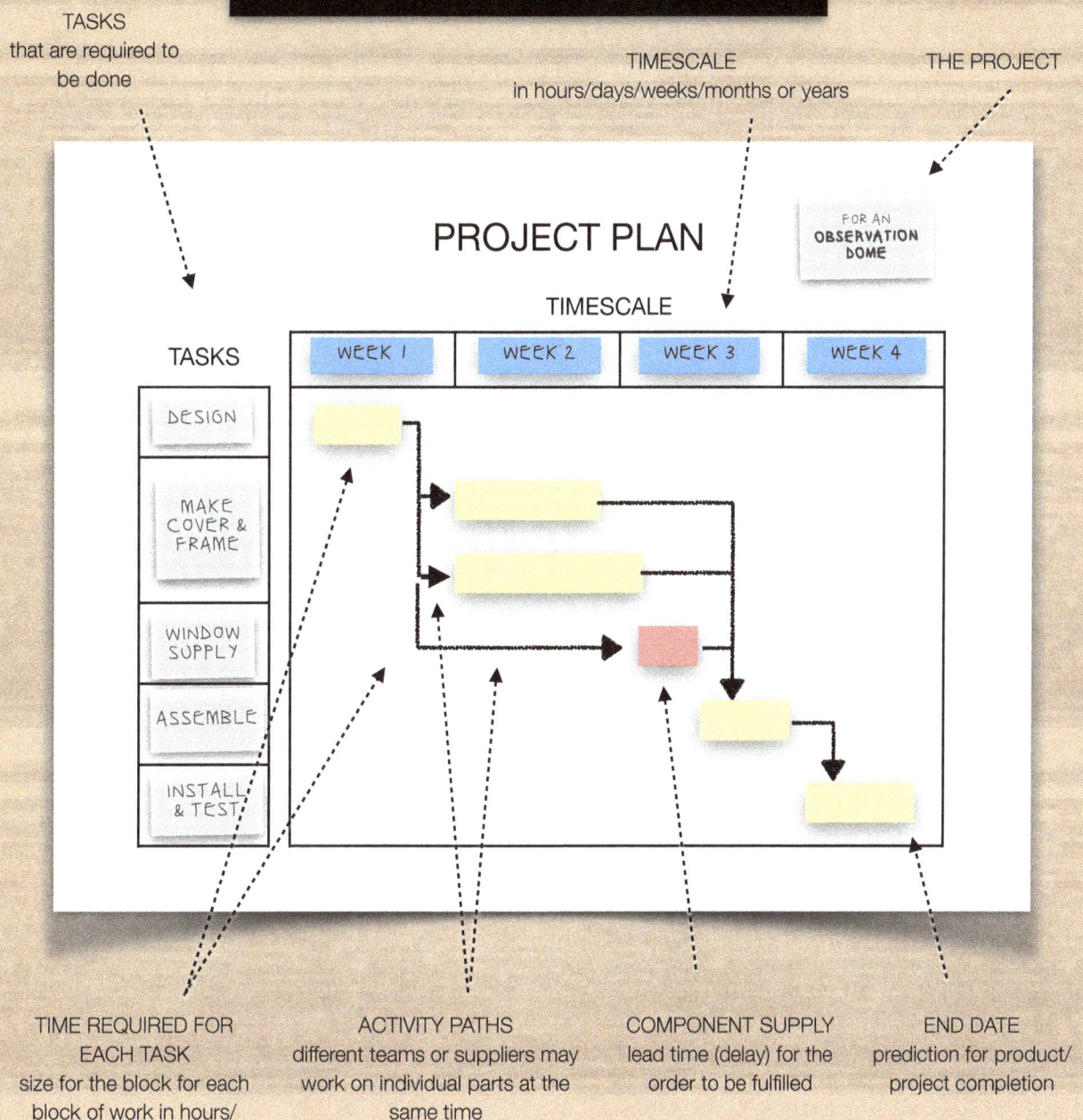

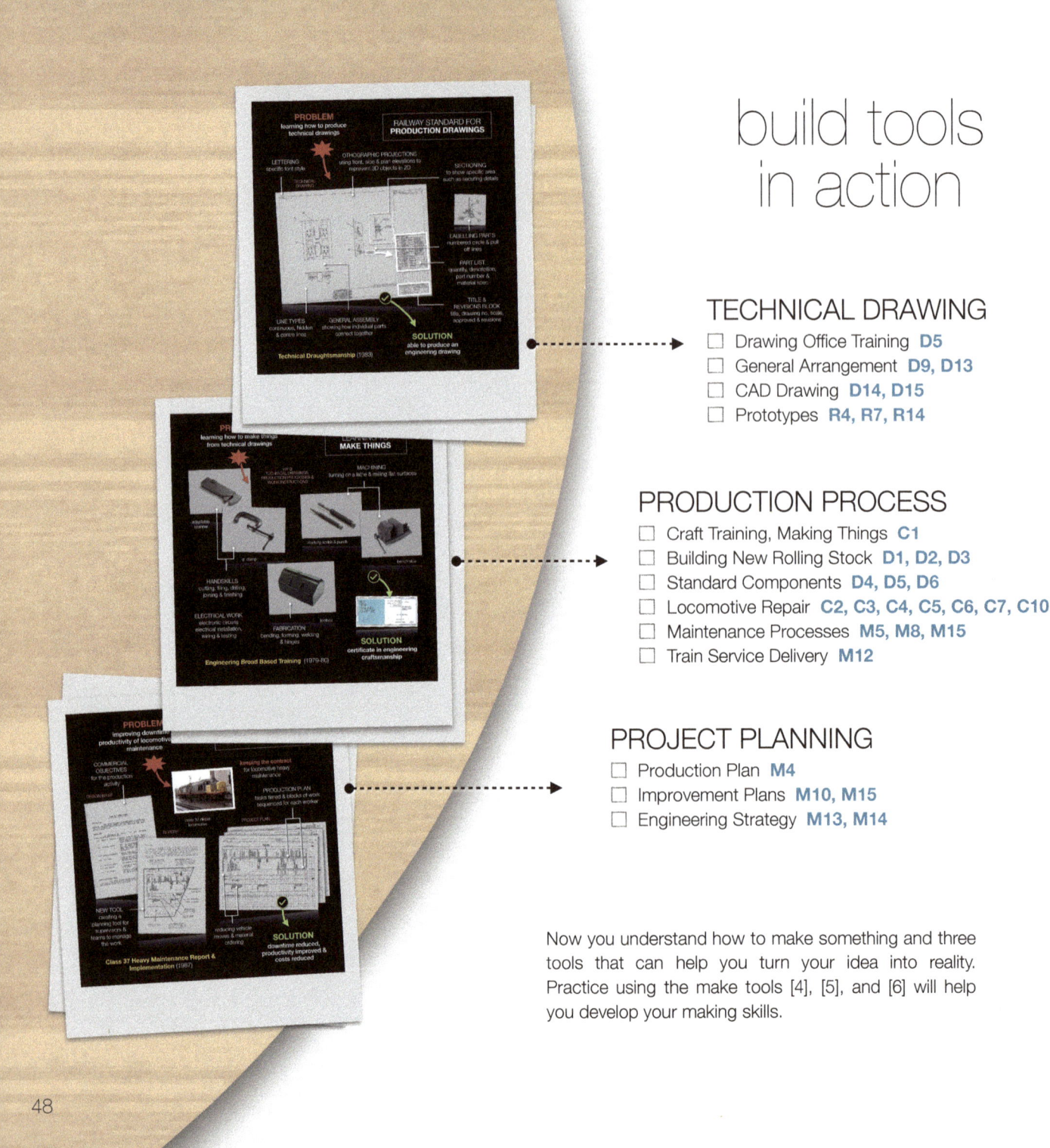

build tools in action

TECHNICAL DRAWING
- ☐ Drawing Office Training **D5**
- ☐ General Arrangement **D9, D13**
- ☐ CAD Drawing **D14, D15**
- ☐ Prototypes **R4, R7, R14**

PRODUCTION PROCESS
- ☐ Craft Training, Making Things **C1**
- ☐ Building New Rolling Stock **D1, D2, D3**
- ☐ Standard Components **D4, D5, D6**
- ☐ Locomotive Repair **C2, C3, C4, C5, C6, C7, C10**
- ☐ Maintenance Processes **M5, M8, M15**
- ☐ Train Service Delivery **M12**

PROJECT PLANNING
- ☐ Production Plan **M4**
- ☐ Improvement Plans **M10, M15**
- ☐ Engineering Strategy **M13, M14**

Now you understand how to make something and three tools that can help you turn your idea into reality. Practice using the make tools [4], [5], and [6] will help you develop your making skills.

3

use & maintain

use & maintain
skills & tools

The word maintain means to keep something in good condition by checking and repairing. In this final stage, we will look at how to prove that the idea is useful, desirable, maintainable, and ideally sustainable.

To maintain their successful solutions engineers answer three key questions :

how well does it work ?
how to use & maintain it ?
how to sell it ?

To determine if the design brief has been met we do **product testing**. These performance tests allow us to demonstrate the product in action and prove that it works or leads to refinements.

To make the product useable we may need to create **user guides** to help the customer learn how to use and look after the product.

Next, we need tell to the world about the product, providing the market with information and persuading them to purchase it. We have a responsibility to ensure that the solutions we develop have a positive **impact on society**, the individual, and the environment.

the maintain stage

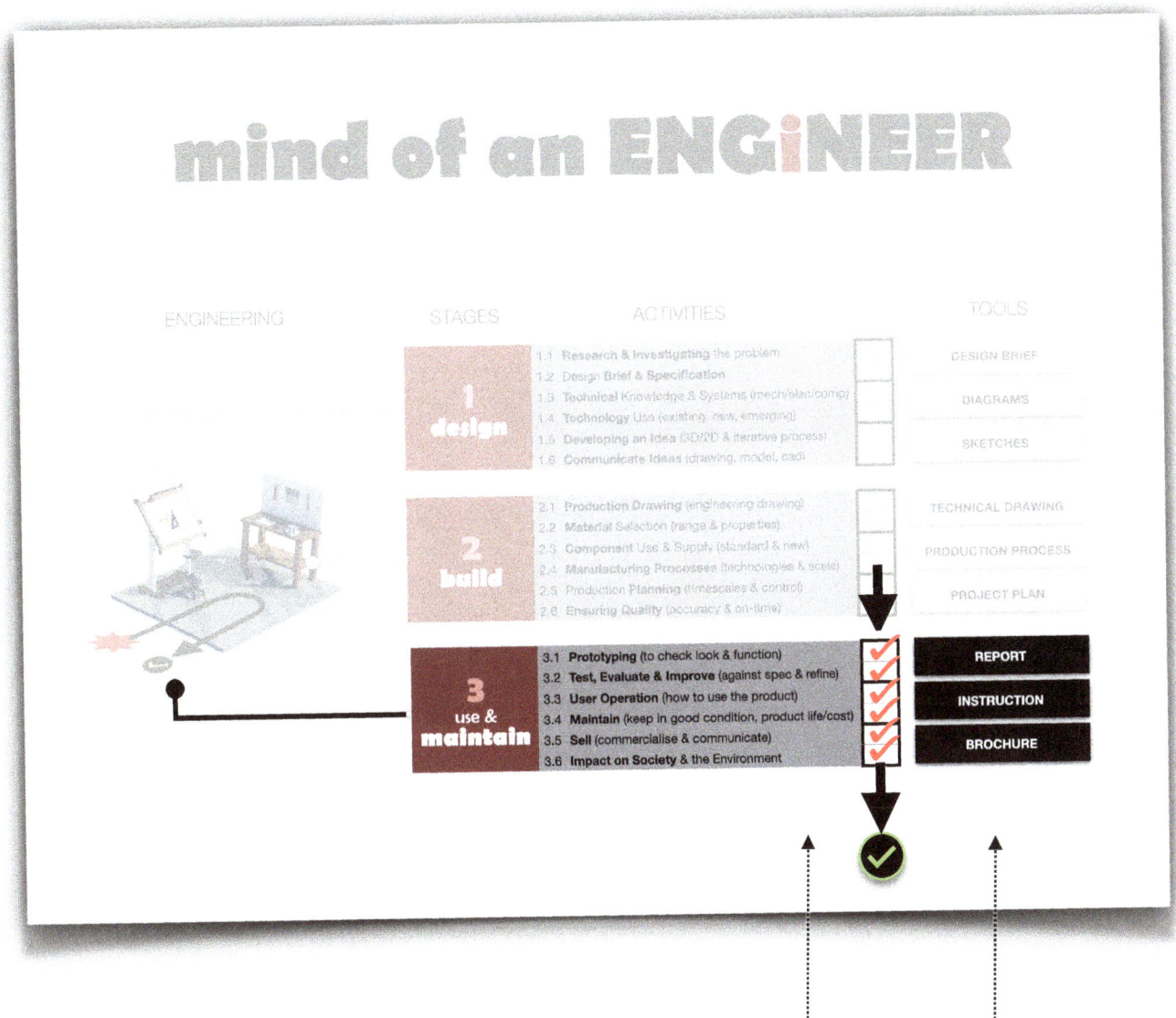

MAINTAIN **ACTIVITIES** & **TOOLS**

how well does it work

... how to use a document & images to answer the QUESTION

HOW WELL DOES IT WORK ?

They design an experiment to test the idea, then conclude and **report** the findings. This often means going back to the drawing board.

| 3.1 | **Prototyping** (to check look & function) |

Prototypes are full-sized, actual versions of an intended product or system, they help us to check how something will look and function. Making prototypes ensures that problems with a design are found early before too much time and money are spent on materials and manufacturing. They can also be presented to clients to gain their feedback.

| 3.2 | **Test, Evaluate & Improve** (against spec & refine) |

Evaluation helps us learn about how well the solution meets the needs of the specification and the client. Evaluation should result in improvements and refinements to the design. It is an ongoing process, all the way to final product manufacture.

REPORT

REPORT
FOR AN OBSERVATION DOME

INTRODUCTION & REQUIREMENTS

- ☑ Simple in design & easy to use
- ☑ Allow full viewing of pantograph contact strip
- ☑ Cheap to manufacture & easy to install

1. DESIGN

2. MAKE

3. USE

Has been in service for several month with no problems to date, providing a clear & close view of pantograph movement/behaviour

CONCLUSION & RECOMMENDATION

- The dome has proved its usefulness as a tool
- Cost £350/unit
- Justification for application to the Class 86/87 fleet by preventing two major de-wirements

---

INITIAL SPECIFICATION — outlining the problem & the requirements for any solution

THE DESIGN — describing the designs & why this option was chosen

EVIDENCE & PHOTOS — showing the final product & design features

TESTING & IMPROVEMENTS — feedback from product testing ie customer reviews & identifying any potential improvements

WHAT NEXT? — make recommendations & outline the next steps ie justifying that it should be commercialised

how to use & maintain it

... how to use a document & images to answer the QUESTION

HOW TO USE & MAINTAIN IT ?

After successful product testing, we must think about helping users. **Instructions** are often produced to explain how to use and maintain the product.

3.3 User Operation (how to use the product)

The owner of a product will often need some instruction on how to use/or assemble, and install the product. For complex devices, owner manuals or quick-start guides will be needed.

3.4 Maintain (keep in good condition, product life/cost)

Over time the performance of a product may alter and will need to be restored. maintenance is the process of keeping something in existence and must be thought about in the design. Activities such as tests, replacements, and repairs need to be defined to keep something serviceable.

INSTRUCTION

INSTRUCTION

FOR AN OBSERVATION DOME

IMPLEMENT : INSTALL

- ☐ Remove existing roof hatch by drilling out rivets
- ☐ Using the dome as a template drill new securing holes
- ☐ Apply sealing compound to the rubber hatch gasket
- ☐ Bolt the dome in place

INSTALLATION INSTRUCTIONS — how to set up or fit the product

MAINTENANCE

3 MONTHLY CHECK
- ☐ Clean the window externally
- ☐ Check the window cover seal
- ☐ Check for any damage & repair
- ☐ Check cover operation

VISUALISED INSTRUCTIONS — showing what things look should like

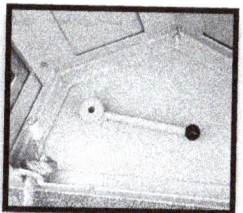

MAINTENANCE SCHEDULE — how & when to check to ensure that the product continues to perform correctly.

INFORMATION

PARTS & MATERIAL
- Window : Triplex ATDS - 2829

REFERENCE INFORMATION — other information that might be useful to know

how to sell it

Now we need to introduce our solution to the market. Creating a **brochure** lets us promote and sell the product.

| 3.5 | **Sell** (commercialise & communicate) |

Products have been created because they solve a problem and benefit the user (society). Marketing helps us explain why your solution is better than other products, then attract potential buyers and influence them to buy.

| 3.6 | **Impact on Society** & the Environment |

Products will affect on wider society, these can be positive and negative. Resource consumption, the sourcing of raw materials, and un-recyclable products impact the future of the planet. Sustainability and thinking about the whole life of a product are important for the environment, society, and the buyer.

... how to use a document & images to answer the QUESTION

HOW TO SELL IT ?

BROCHURE

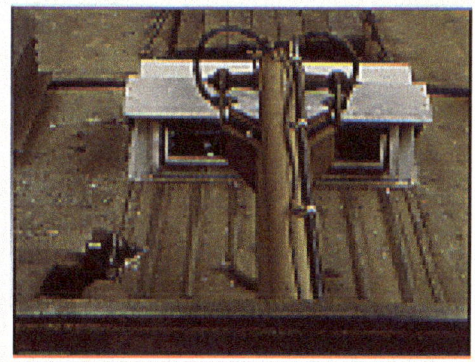

PRODUCT NAME
naming the product & giving a quick description

COMPANY & BRAND
brochure design to reflect the branding of the company & include any logos

PRODUCT DESCRIPTION & BENEFITS
what is the problem it solves, why you should buy the product & the impact it could make

IMAGES OF THE PRODUCT
showing what it looks like & it being used

CONTACT DETAILS
how to find out more

maintain tools in action

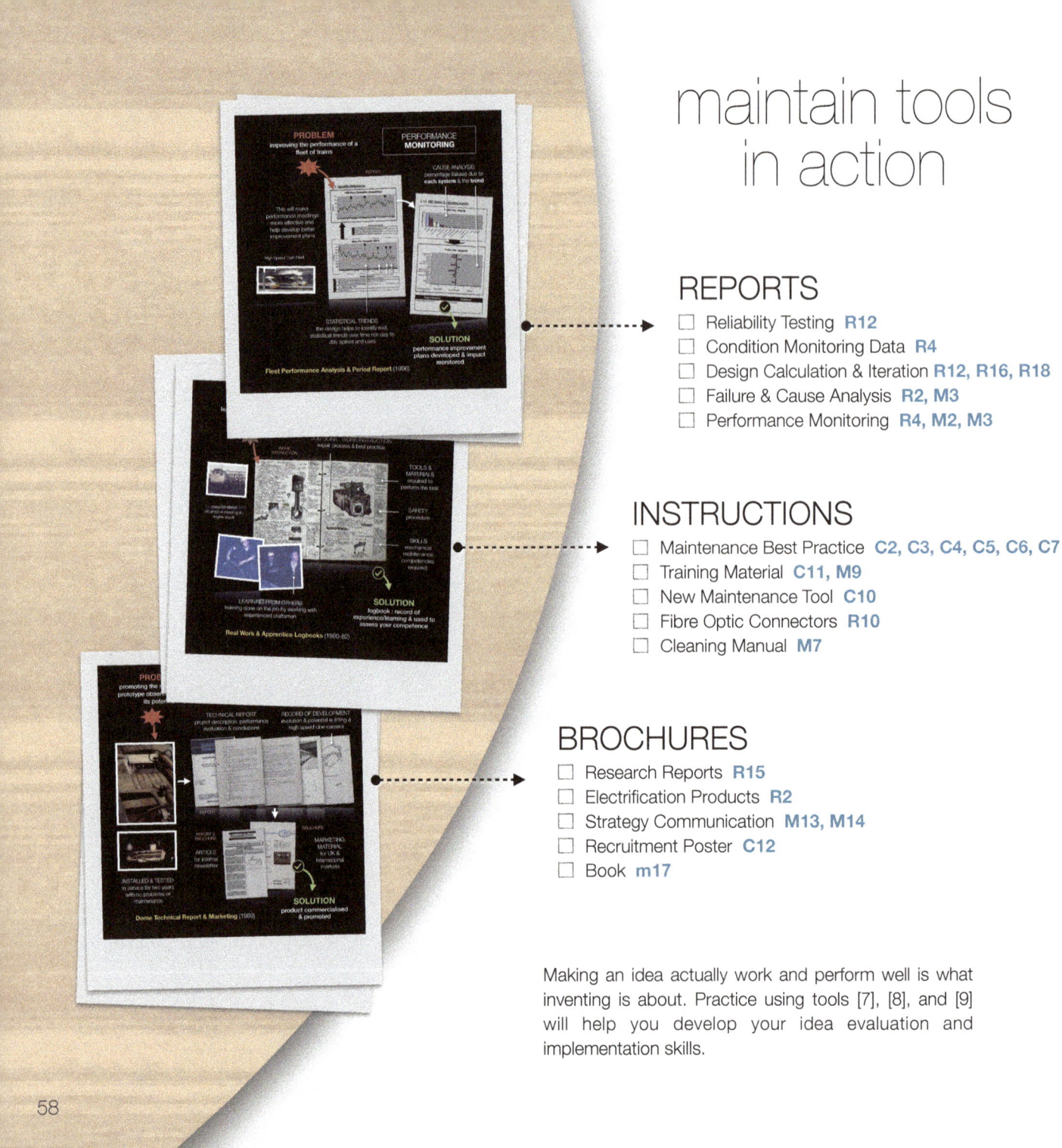

REPORTS
- ☐ Reliability Testing **R12**
- ☐ Condition Monitoring Data **R4**
- ☐ Design Calculation & Iteration **R12, R16, R18**
- ☐ Failure & Cause Analysis **R2, M3**
- ☐ Performance Monitoring **R4, M2, M3**

INSTRUCTIONS
- ☐ Maintenance Best Practice **C2, C3, C4, C5, C6, C7**
- ☐ Training Material **C11, M9**
- ☐ New Maintenance Tool **C10**
- ☐ Fibre Optic Connectors **R10**
- ☐ Cleaning Manual **M7**

BROCHURES
- ☐ Research Reports **R15**
- ☐ Electrification Products **R2**
- ☐ Strategy Communication **M13, M14**
- ☐ Recruitment Poster **C12**
- ☐ Book **m17**

Making an idea actually work and perform well is what inventing is about. Practice using tools [7], [8], and [9] will help you develop your idea evaluation and implementation skills.

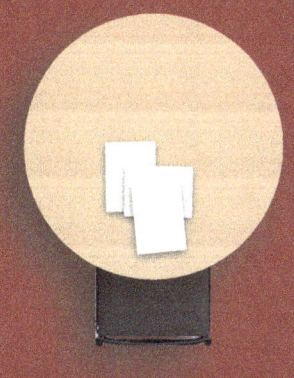

engineering
projects

innovation or improvement

Now you've got a set of tools to help you solve problems quickly, effectively, and successfully. To improve the world you have two choices, to improve an existing solution (build on past successes) or innovate to create something fundamentally new.

We know that the end-to-end process or product development process goes from idea to drawing, drawing to thing, and thing to product. All these activities are normally managed through the creation of a project, and a plan produced. The opposite page shows the typical journey of an idea and how a project could be managed.

the journey of an idea & planning a project

It is not a straight line but is a dynamic and iterative one, it is an experimental approach where ideas will be tested and refined many times as we work towards a solution. To progress faster when we face up to failure and learn from it.

Doing new things, and inventing the 'what next' is hard to do because you are doing something that has never been done before. Some solutions are quick and require only a little research and the selection of the appropriate part. Other cases require large teams and extensive testing of multiple prototypes. Innovation is hard to do, high risk, and takes a lot of time.

planning a project

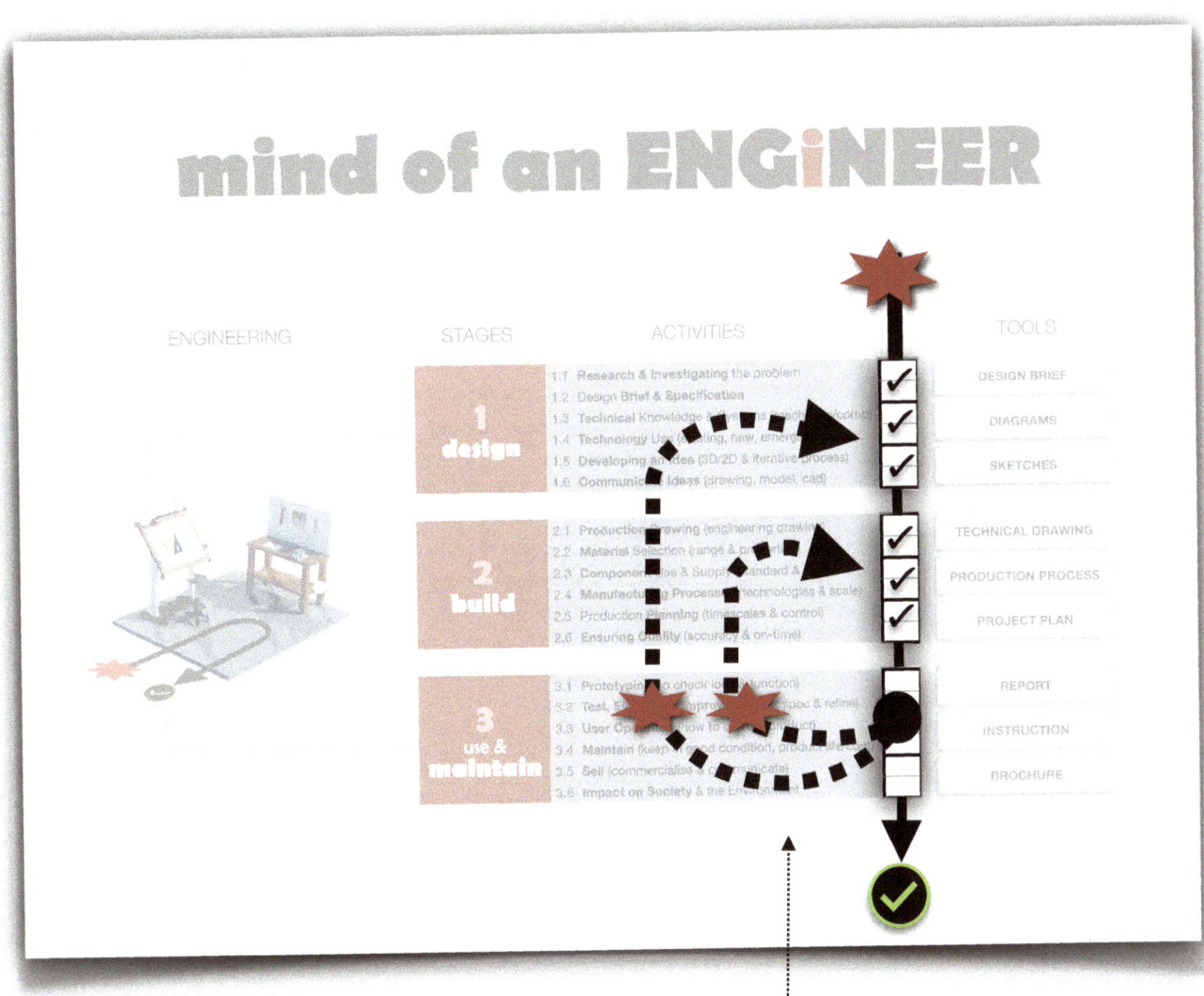

RE-DESIGN or RE-BUILD
LOOPS

collaboration & teamwork

Few things are by a single individual, engineers seldom work in isolation and most tasks they undertake require a range of expertise in designing, building, and maintaining. A project is commonly delivered by a team, we start by understanding our capability and the specialist support we will need.

The whole process of taking something that 'might work' through many stages of development will require the right mix of people with the right skills. This normally means contributions from other functions or specialists, so we will have many engineers working together on single projects, and being able to work with other experts from other areas is important.

undertaking **a group project**

The bringing together of these skills, expertise, and experience is often managed through the creation of a project team. The project team will need to be designed and managed for the unique abilities of all team members to be used and bring about a successful solution.

This group of diverse people will need to work collaboratively, to share collective insights, ideas, and issues to build great solutions. It will take teamwork, communication, and planning to do this. Effective leadership is needed. The management of people is often the most challenging, which makes it necessary to appoint a project manager.

the project team

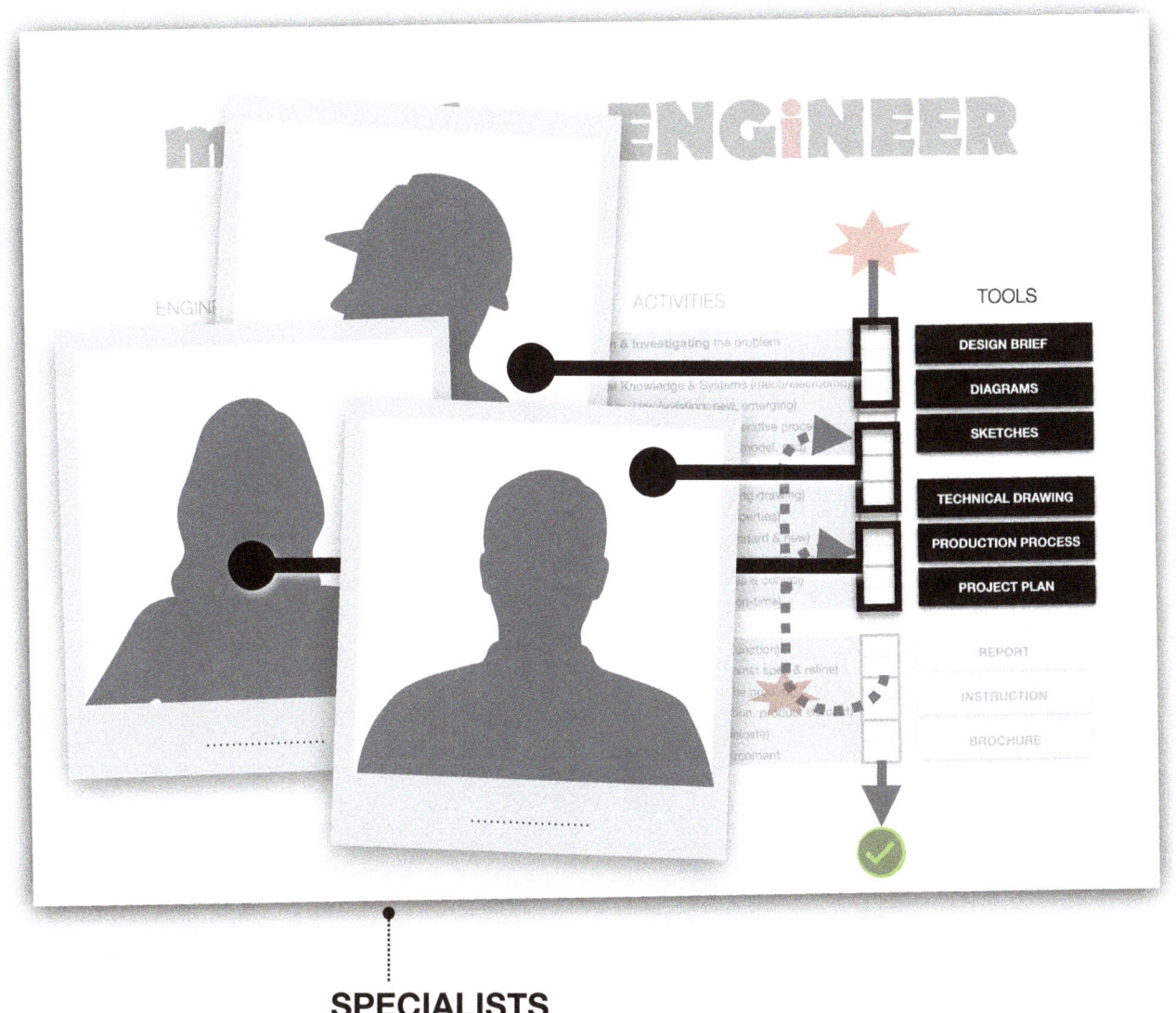

SPECIALISTS
SKILLS & ROLES REQUIRED

developing ideas

Our initial ideas can be brought to life by using our first four tools. These are used in three basic ways : to visualise the design idea, to communicate the idea so others can evaluate it, and to document the design so the product or process can be reliably reproduced and maintained.

Putting things on paper or a computer screen gives you the freedom to think and experiment with your ideas. Having gathered together our thoughts and visualised our ideas, we need to develop them into an initial 3-dimensional sketch or model.

bring to life new concepts

The design is documented so that will be understood by the manufacturer using drawing conventions. This means developing 3-dimensional models into 2-dimensional detailed technical drawings, these show up problems not apparent in the sketches. Through technical drawings the manufacture of equipment can be broken down be made and/or contracted, and sub-contracted.

Technical drawings allow prototypes to be made and improved on quickly. They are living documents that can be continually updated and are a live record of the development journey. The four tools and drawings allow the often invisible and non-linear act of Innovation to be seen and accelerated.

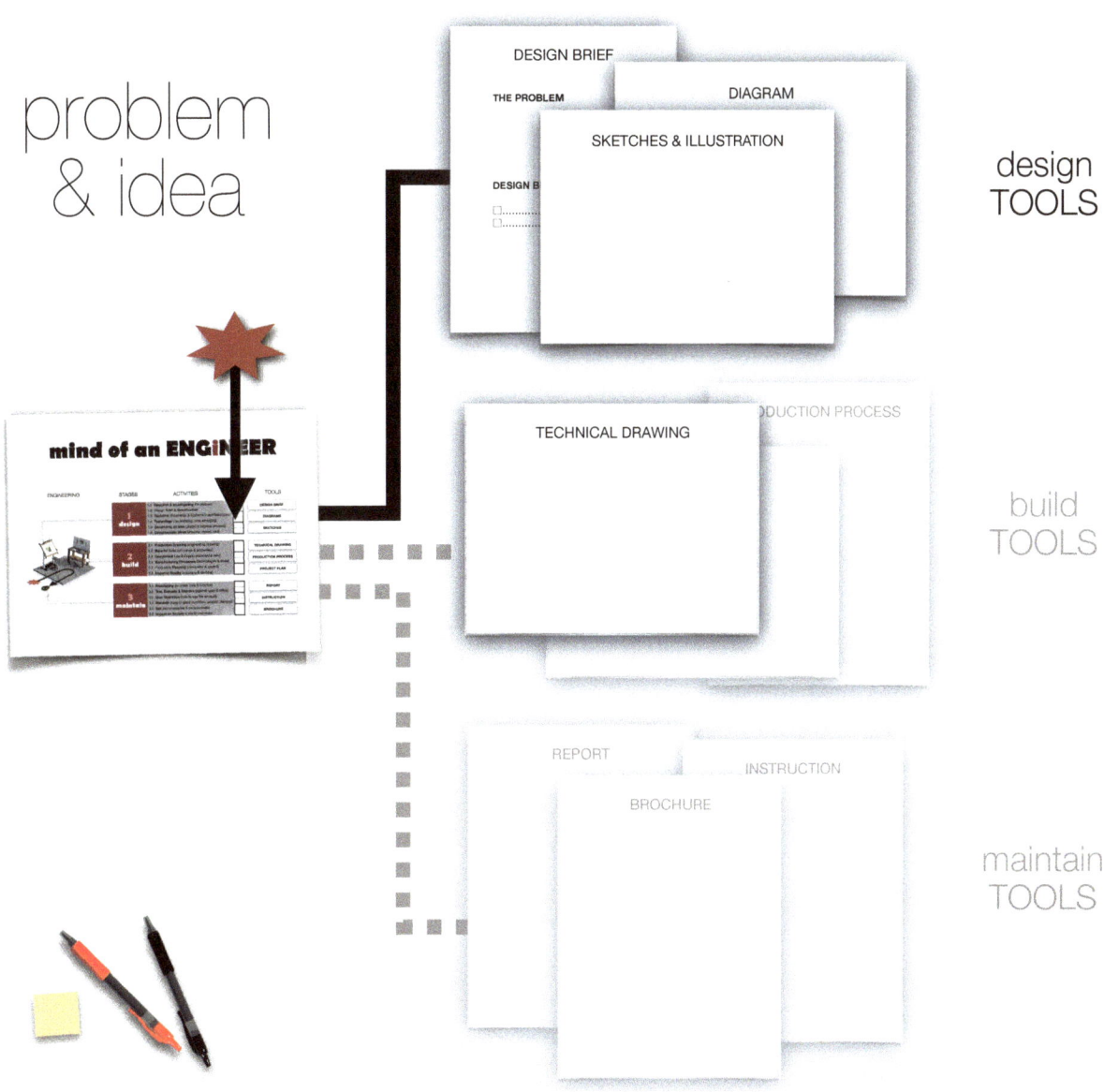

implementing ideas

Once the early concept has proved feasible we need to show that it can be made and quantify the resources needed. Then show how we will manage and organise the creation of the new product we need to define how long it will take to get it to market.

To conclude and justify the project we will need to prepare and produce a report. Attracting support may involve making a pitch and giving a presentation to an audience. To demonstrate that risk is worth the investment, the presentation needs to educate newcomers, define the support required, and build a business case.

turning the idea **into reality**

Once approved the solution needs to be introduced to the market. Creating a brochure lets us promote and sell the product, and creating instructions tells the user how they can use it. Marketing helps us explain why your solution is better than other products, then attract potential buyers and influence them to buy.

Products have been created because they solve a problem and benefit the user and society. User feedback then provides validation for a successful design and will hopefully result in commercial success. But user's needs will change over time so even after the solution is in full production, there will be continual development taking place.

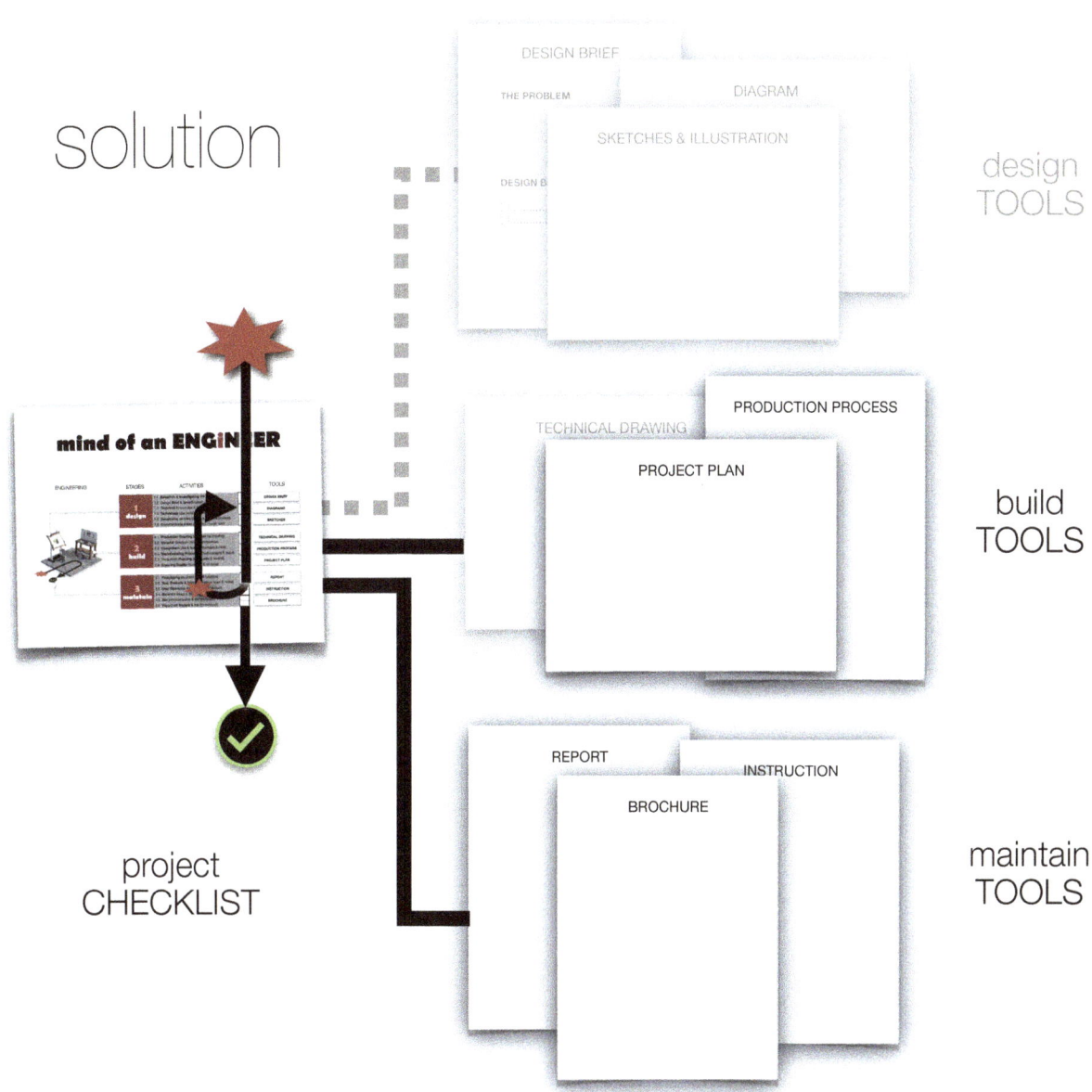

have a go

You can start a project yourself, from concept to solution now. If you want, you can make the world better and find a way to solve problems sustainably. Although the focus is on products that are physical and engineered, the method applies well to a broad range of products or services.

REAL PROJECTS

It is important to remember that no engineering problem is ever completely solved to everyone's satisfaction and there is always room for improvement in the real world. But engineers realise that they must at some point curtail design and begin to manufacture or build. Part [ii] of this book will show examples of some real projects that show the process of designing, building, and maintaining in action.

part ii

engineers IN ACTION

ENGINEERS IN ACTION

In this part of the book, you will see how engineers solve real problems and how they develop as problem solvers. The six chapters give you an insight into the mind of an engineer and engineering practice :

- ☐ real JOBS (x4) & workplace realities
- ☐ how engineers DEVELOP
- ☐ real PROBLEMS (x66) & PROJECTS

HELPING YOU :

- ☐ career planning
- ☐ professional development
- ☐ solve non-engineering problems

The chapters help you develop yourself as a problem solver for education (pages 210 & 212) and help with professional development (page 214).

what engineers DO

types of **problem**

Engineering is a broad discipline and engineers work in quite a diverse range of industries. Each industry will have a slightly different mix of challenges, but there are four common types of problems that they are called to work on. The problems increase in difficulty from making and fixing things, to designing and then to inventing.

The chart opposite shows the relationship between the increasing complexity of the problems being solved and the competence required. Some problems such as making and fixing things are well defined and proven techniques and procedures have been developed to solve them, whilst more complex problems may have no obvious solution and require originality.

what type of **problem** : well-known or complex problems

The vertical scale shows how we need to increase our knowledge, understanding, and skills to solve more complex problems. The engineering profession has established a set of competence standards for the different levels, that every engineer has to achieve and demonstrate.

The four main traditions or engineering disciplines are mechanical, electrical, civil, and chemical engineering, with the more recent addition of digital/software engineering. The focus of the book is on mechanical and electrical engineering, with a little bit of digital.

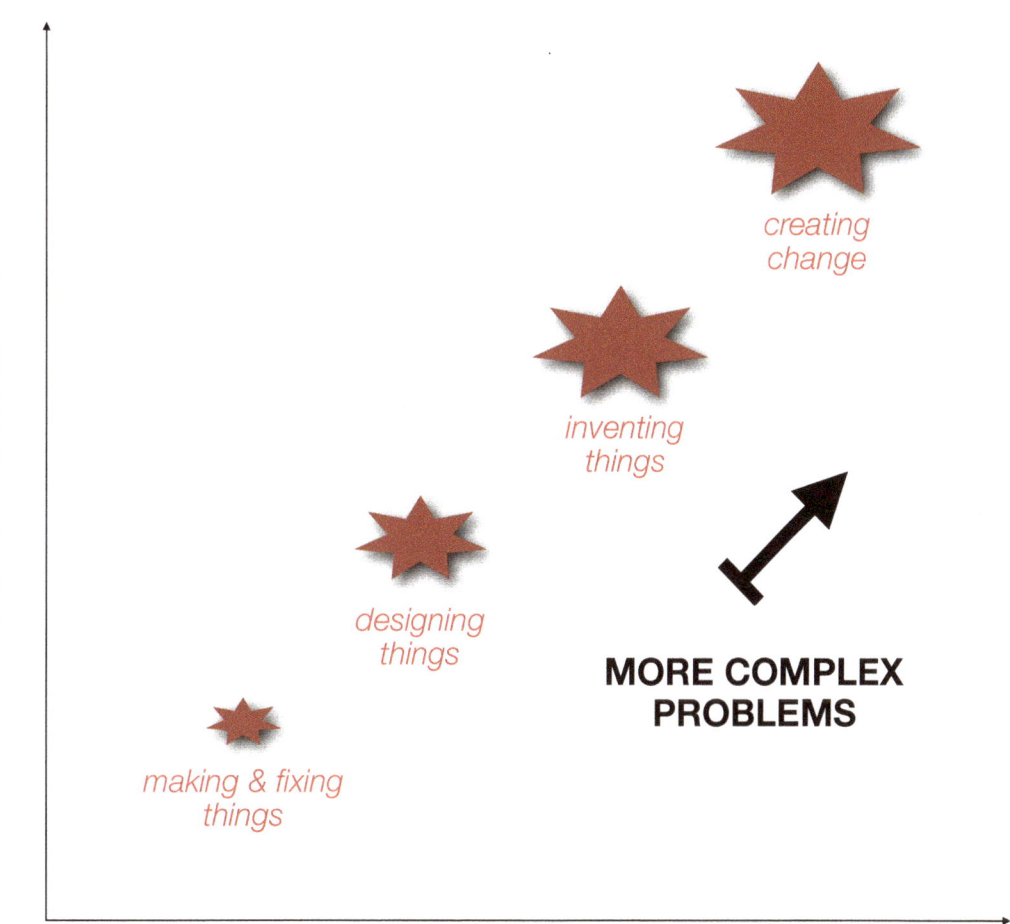

engineering **jobs**

Different problems need different problem solvers and the different engineering jobs will have a specific mix of design, make, and use skills. There are many kinds of engineers, but all share the ideas and methods introduced in this book.

We will look at how different four jobs at different levels and in different departments use the knowledge and tools from part (1). For each of the four jobs, the role they play in the day-to-day work they do in the engineering field

types of **problem solvers** :
craftsman or engineer

The **craftsman** possesses the manual skills necessary to produce the components specified by engineers skilled trades possess the skills necessary to produce parts and abilities to build and maintain specialised equipment.

What distinguishes the **designer** and **research scientist** from the craftsman is largely the ability to formulate and carry out detailed calculations of forces and deflections, contraptions and flows, voltages and currents, that are required to test a proposed design on paper or computer screen. The ability to predict the performance of a design before it is built and tested.

Success as an engineer will often mean a transition to leadership. The engineering **manager** must learn to work with and through people. Management of people is often the most demanding and most difficult to learn and the most rewarding aspect of it all.

jobs & competence levels

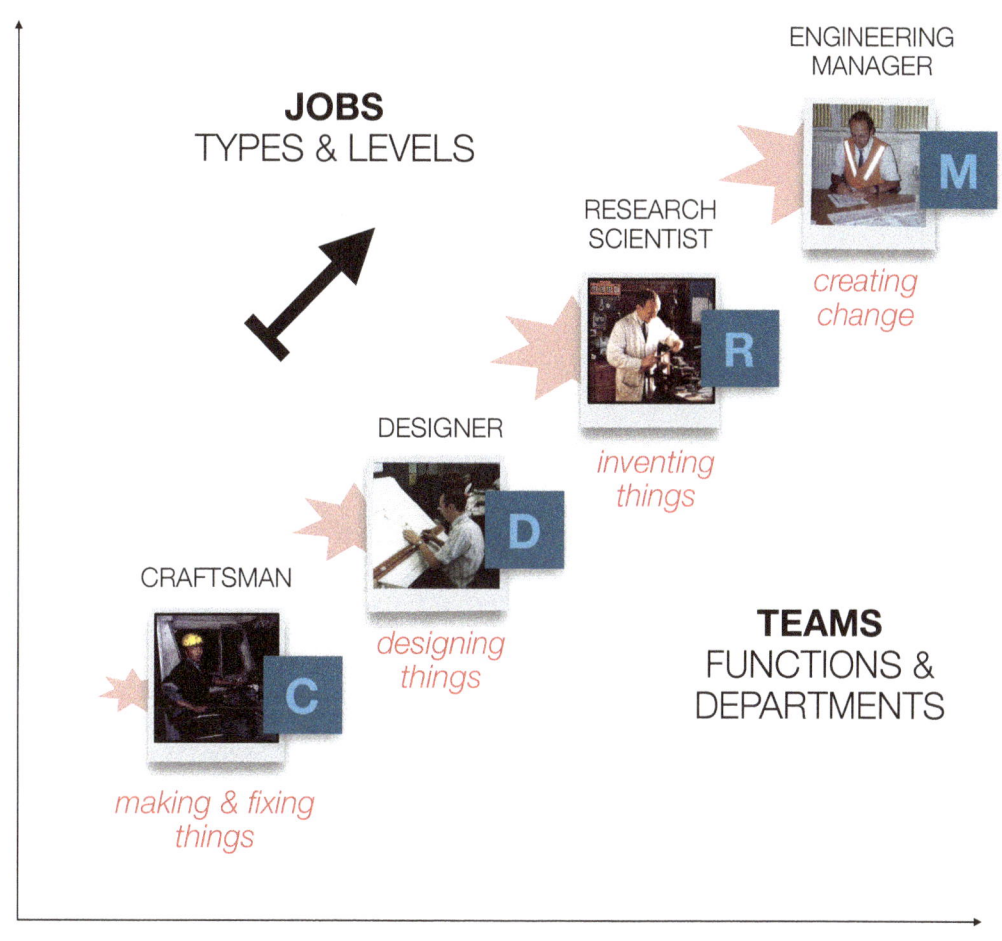

work
& workplaces

Each industry will have a slightly different mix of challenges and therefore skills. I'm going to use my experience in the rail industry, my work, and some of the projects I've been working on to give you an insight into the world of engineers.

take you **inside the workplace**
and into the business

Go Inside **workshops**. depots where trains are maintained and repaired, and see craftsmen tackling electrical and mechanical work. Go inside the heavy engineering workshops where the trains are built.

Go inside the **design** or **drawing office** to see an engineer designing and developing new fleets and improving existing vehicles. See how they use sketches and dimensioned drawings to tell a story and give instructions.

Inside a **research and development** division, to see scientific engineers inventing new products and new processes that have their roots in research and follow the path from laboratory idea, through prototype production, manufacturing, and market introduction.

Visit the **manager's office** to see how they work with people on developing and implementing business improvements. Find out how these tools can be used to develop new ways of working, and employees are developed into problem solvers.

disciplines & specialities

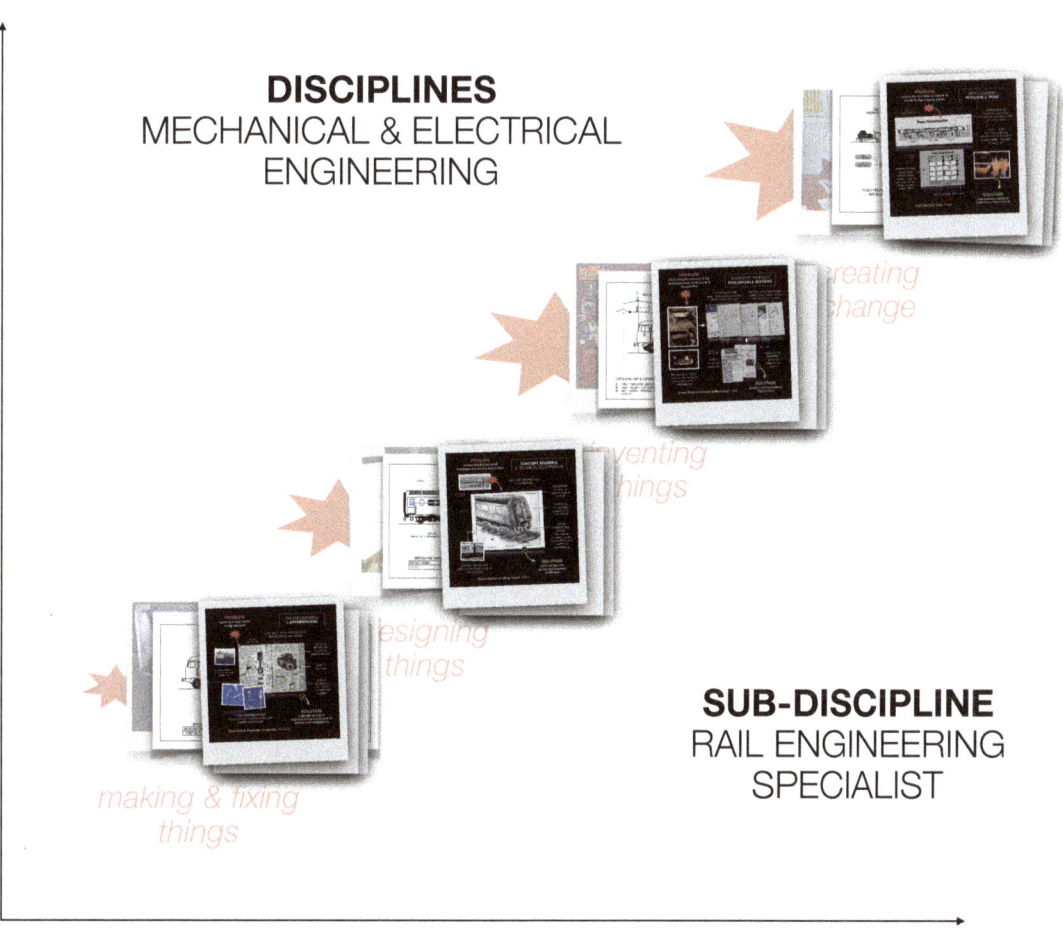

engineering **work**

Engineering work is about solving problems. However many engineering **problems** start by being ill-defined, so engineers must act in the absence of complete knowledge and make judgements under conditions of uncertainty. To help develop a picture of engineering work we will look at how real problems are solved.

Engineering work involves the **integration** of the problem-solving process with technical knowledge. All of this happens invisibly within the mind of a single engineer and between engineers, so I have used personal notebooks to help you see the thought process used.

> engineering is, at its core,
> **problem-solving**

A range of different problems and examples of **real work** (polaroid pages) show the process from problem to solution and how the tools in part [i] are used. Sixty-six real problems will hopefully help you see how they do it and see the tools being used, sometimes in different ways and projects where multiple linked problems have to be solved.

Engineers possess fundamental **technical knowledge**. Different things (cars, trains, planes) need different specialist technical knowledge, knowledge about the specific product and the technologies they use. What engineers do depends on what they know, so for each of the jobs discussed I have introduced some of the specialised knowledge they need to acquire.

insights into engineering practice

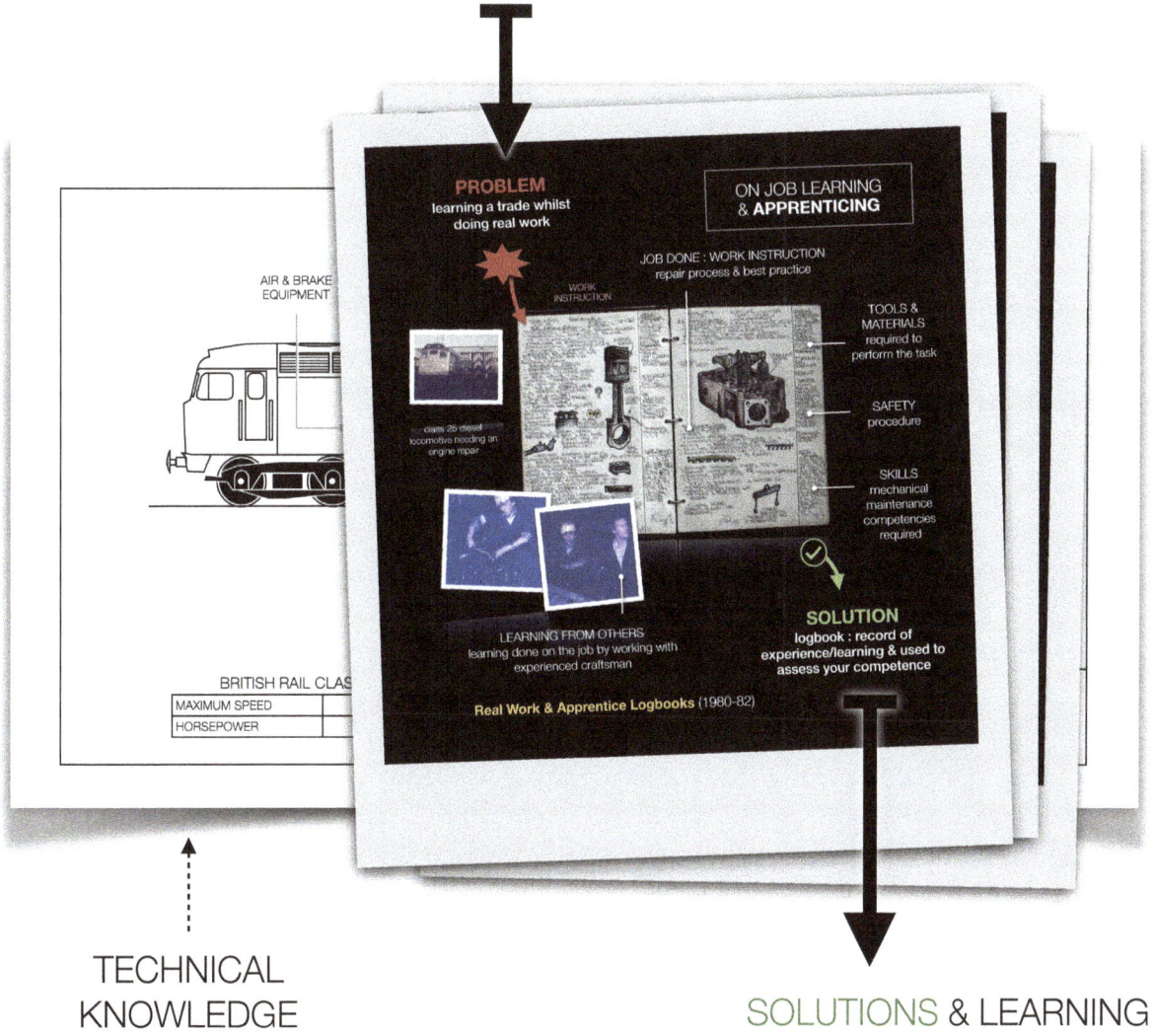

continuous change

The rail industry is a great case study because although railways are simple in concept, vehicles running on track are incredibly complex and technically challenging. Rail vehicles are expensive assets that have to last a long time, so there is a continual search for better and more efficient traction and rolling stock.

constantly evolving & never staying still
new challenges, new technologies
& industry changes

Although the problems and projects shown here are from the rail industry and were done several years ago, the principles and processes are still key to all engineering work today.

The timeframe of the work was between 1980-2000 is also really interesting with the rail transport with new fleets of trains being developed and a time of radical innovation with new technologies emerging. The railways were still nationalised so operated as an integrated whole, today it is a privatised and fragmented industry.

the job of a **craftsman**

MAKING & MAINTAINING THINGS

Modern equipment is complicated, and sometimes delicate and needs expert maintenance by real craftsman. The mind of an engineer sheet opposite summarises what a craftsman does, the type of problems that they solve, and skills, and tools they use.

A craftsman's job was to keep trains and plant equipment in working good condition by applying engineering concepts and carrying out the following tasks :

fault finding & rectification
(by understanding the **design**)

build or repair things,

maintain things to keep them
in existence & operating

The problems craftsmen solve are **known problems** with existing products or things. However because rail vehicles are made up of lots of complex systems and thousands of parts, the challenge is to diagnose the fault.

Technical understanding is needed to identify causes. **Schematics diagrams** and **technical drawings** give them this knowledge.

They then select and apply proven techniques, procedures & methods to solve practical engineering problems. These best practices are stored in **work instructions** so that tasks can be done safely, efficiently, and in a controlled, auditable way.

job profile

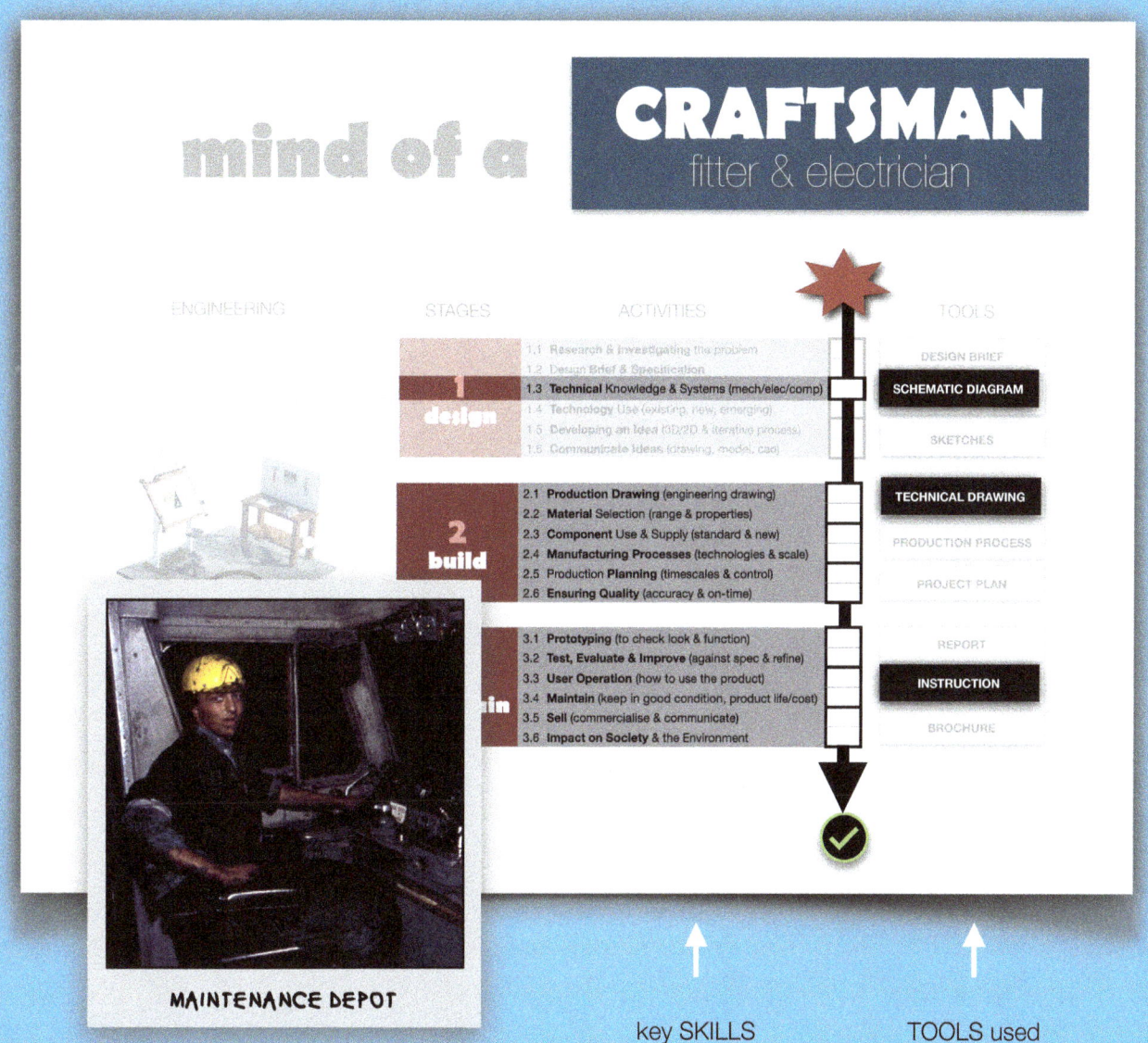

key SKILLS   TOOLS used

inside the maintenance depot

THE MAINTENANCE DEPOT

A railway needs to be able to maintain diesel and electric locomotives, rolling stock, and all types of plant & machinery to operate.

Railway craftsmen would need to be able to tackle any type of work and would be trained to be competent in both electrical and mechanical engineering. This gave the employer the flexibility and productivity they required.

CRAFTSMAN : TECHNICIAN

To become a craftsman I would undertake a 4 year apprenticeship. Craft knowledge and skills would be learned on the job and follow a development path that would help the trainee achieve the required industry standards.

The apprenticeship would start with a period in a **craft training** workshop where over a year you would learn how to make things.

This was then followed by practical training and gaining experience in railway maintenance depots. Carrying out **electrical & mechanical maintenance** work, **fault finding**, and recording achievements in a logbook to show evidence of competence.

The experience and my logbooks would help open up other **career opportunities.** I would later work in technical support, in a technical training team and be asked to create a national apprenticeship recruitment poster.

workplace

Maintenance Depots

Area Maintenance Depots, Wolverhampton
Responsible for the Maintenance of Traction, Rolling Stock, Plant & Stations in the West Midlands

Carrying out Electrical & Mechanical Maintenance on a range of equipment:
- Diesel Locomotives
- Electric Traction
- Carriage & Wagon
- Cranes, Mobile & Gantry
- Track Machines
- Stations & Equipment
- New Electrical Installations : Stations

Bescot Traction Maintenance Depot

Crane, Steel Yard Wednesbury

BRITISH RAIL

British Rail was amongst the worlds most modern railway systems, operating passenger & freight rail transport. Services on the network were provided by five geographic regions, Southern, Western, Eastern, London Midland and Scottish.

MAINTENANCE DEPOTS

Working at Bescot Traction, Wolverhampton Plant & Oxley Carriage Maintenance Depots on the London Midland Region of BR

Apprentice Engineer

Craft Apprenticeship
A four year apprenticeship, starting with a period in a training workshop (Bilston) followed by practical training and controlled experience in maintenance depots

Qualification
- Certificate (J1) in Electrical Maintenance & Installation
- Certificate (J2) in Technical Maintenance & Installation
- Broad Base Training & Electrical/Electronic Engineering (First Year)
- Further Education : TEC Certificate in Electrical Engineering

Apprentice Logbooks (x70 logs)

Class 25 : Cylinder Head Change

Class 47 : A Exam & Fault Finding

MAINTENANCE TEAMS

Teams of 5-10 craftsmen work on shifts (day, night etc) & carry out a mixture of electrical & mechanical maintenance work

QUALIFICATIONS

Certificate in Engineering Craftsmanship for Electrical & Mechanical Maintenance and BTEC (level 3) Certificate in Electrical & Electronic Engineering

maintenance **technical** knowledge

RANGE OF RAILWAY EQUIPMENT

Maintenance work requires knowledge of both electrical and mechanical engineering and systems. These universal engineering principles & an understanding of systems help the engineer work on a wider variety of rail equipment.

HOW A LOCOMOTIVE WORKS

There are two types of Locomotive diesel and electric. A diesel locomotive is a self-powered vehicle that uses electricity to drive forward motion, they function by :

- Batteries are used to start the Engine
- The Diesel Engine drives a Generator
- The Generator makes Electricity
- The electrical energy powers Traction Motors
- The motors turn the Wheels
- And controlled by Electrical & Electronics

The generator also powers auxiliary equipment for the engine (cooling, fuel & oil systems) and train braking, heating, lighting & electrical power.

Rail vehicles are mounted on bogies that are articulated to allow cornering and carry the wheel-sets & suspension systems.

railway equipment

KNOWLEDGE OF ELECTRICAL & MECHANICAL SYSTEMS

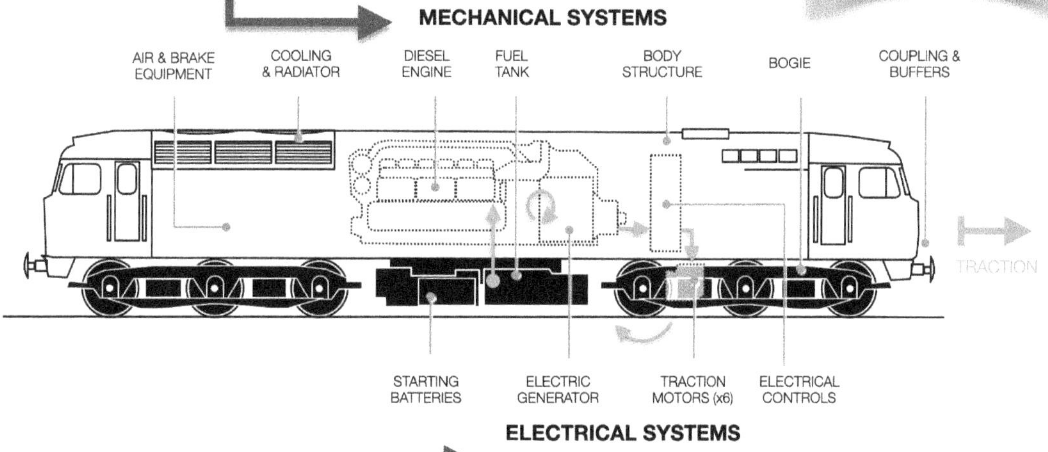

MECHANICAL SYSTEMS

- AIR & BRAKE EQUIPMENT
- COOLING & RADIATOR
- DIESEL ENGINE
- FUEL TANK
- BODY STRUCTURE
- BOGIE
- COUPLING & BUFFERS

TRACTION

- STARTING BATTERIES
- ELECTRIC GENERATOR
- TRACTION MOTORS (x6)
- ELECTRICAL CONTROLS

ELECTRICAL SYSTEMS

BRITISH RAIL CLASS 47

MAXIMUM SPEED	95mph
HORSEPOWER	2400 bhp

OPERATIONAL USES
- PASSENGER (& SLEEPERS)
- FREIGHT

Title
CLASS 47 DIESEL LOCOMOTIVE

VARIOUS RAILWAY EQUIPMENT
including diesel & electric locomotives, rolling stock and plant & machinery

craft apprenticeship

The path below follows the development over three to four years of a craftsman, to become a qualified practical problem solver in both electrical and mechanical maintenance. And how the experience led to other career opportunities…

ELECTRICAL MAINTENANCE

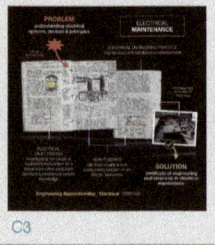

C3

electrical engineering practice

how to make things?

CRAFT TRAINING

how to fix electrical things

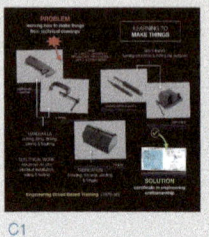

C1

C2

one year in a training workshop learning how to make things by hand or with machines, and using engineering drawings

on-job learning & logbooks

how to fix mechanical things

MECHANICAL MAINTENANCE

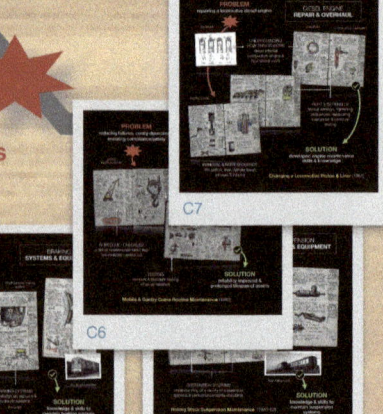

C7

C6

C4 C5

mechanical engineering practice

APPRENTICE LOGBOOK

A logbook would be used to record the knowledge and skills you achieved during your apprenticeship. Over two years 80 log entries would need to be completed to show a range of jobs/tasks and show evidence that you met industry competence standards.

DUAL TRAINED FITTER & ELECTRICIAN

Being dual-trained in electrical & mechanical engineering meant that a craftsman could understand how every system and component on a locomotive, coach, crane, piece of plant, etc worked. This was key to effective fault finding and rectification. This also created a very flexible and effective workforce for the employer.

PROGRESSION

The knowledge and experience from the apprenticeship would be used to develop career opportunities.

TECHNICAL SUPPORT JOB

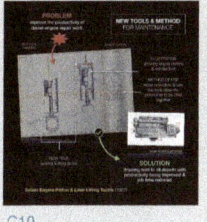

worked in a drawing office & creating instructions for maintenance teams

FAULT FINDING

why isn't it working ?

FAILED LOCOMOTIVE

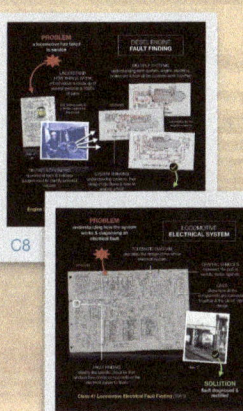

fault diagnosis using schematics & understanding the interaction between mechanical + electrical systems

QUALIFIED CRAFTSMAN

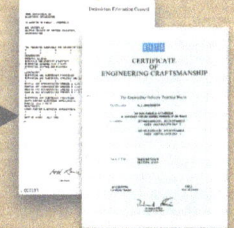

achieved nationally recognised awards, certificate in craftsmanship & electrical engineering

TECHNICAL TRAINING JOB

developing training courses & materials for the introduction of a new fleet of trains

GRAPHIC DESIGN WORK

designed a national recruitment poster, in my spare time

DOING REAL WORK

Learning by doing real work and learning from others who have solved the problems before, makes an apprenticeship an effective system for learning & development.

OFF JOB - ACADEMIC STUDY

Workplace training was combined with study, to give a knowledge and understanding of engineering, science, and mathematical principles.

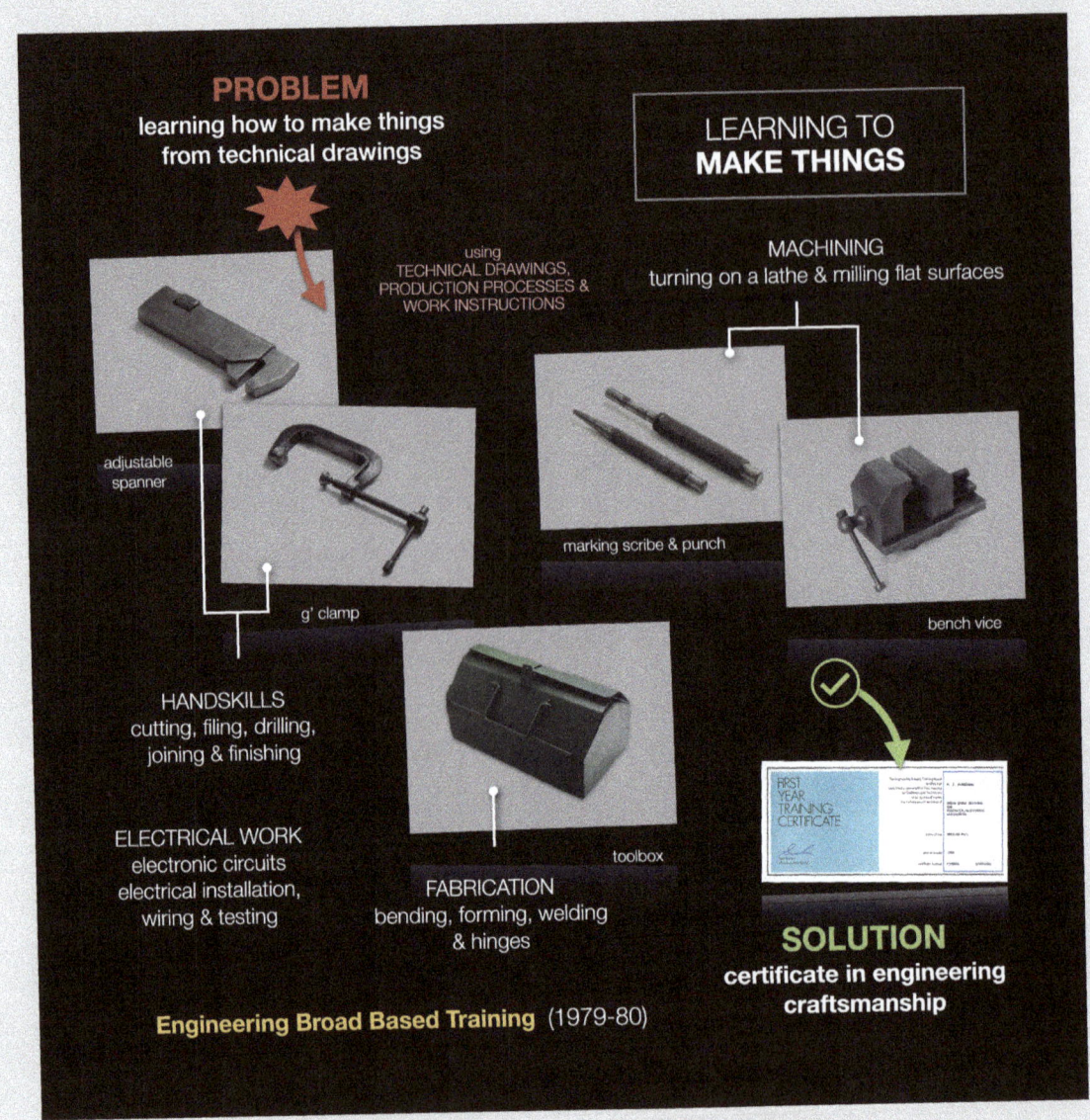

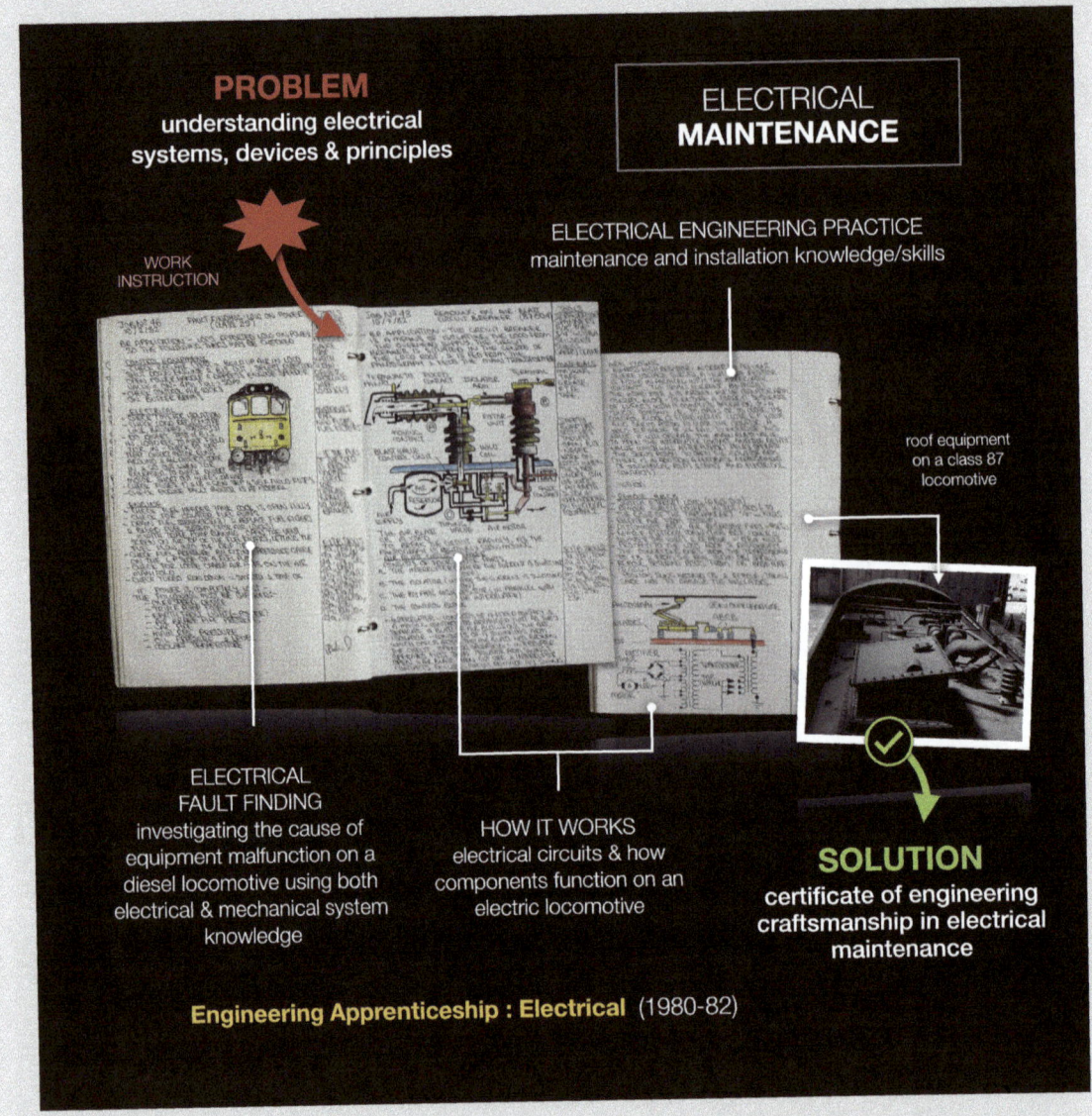

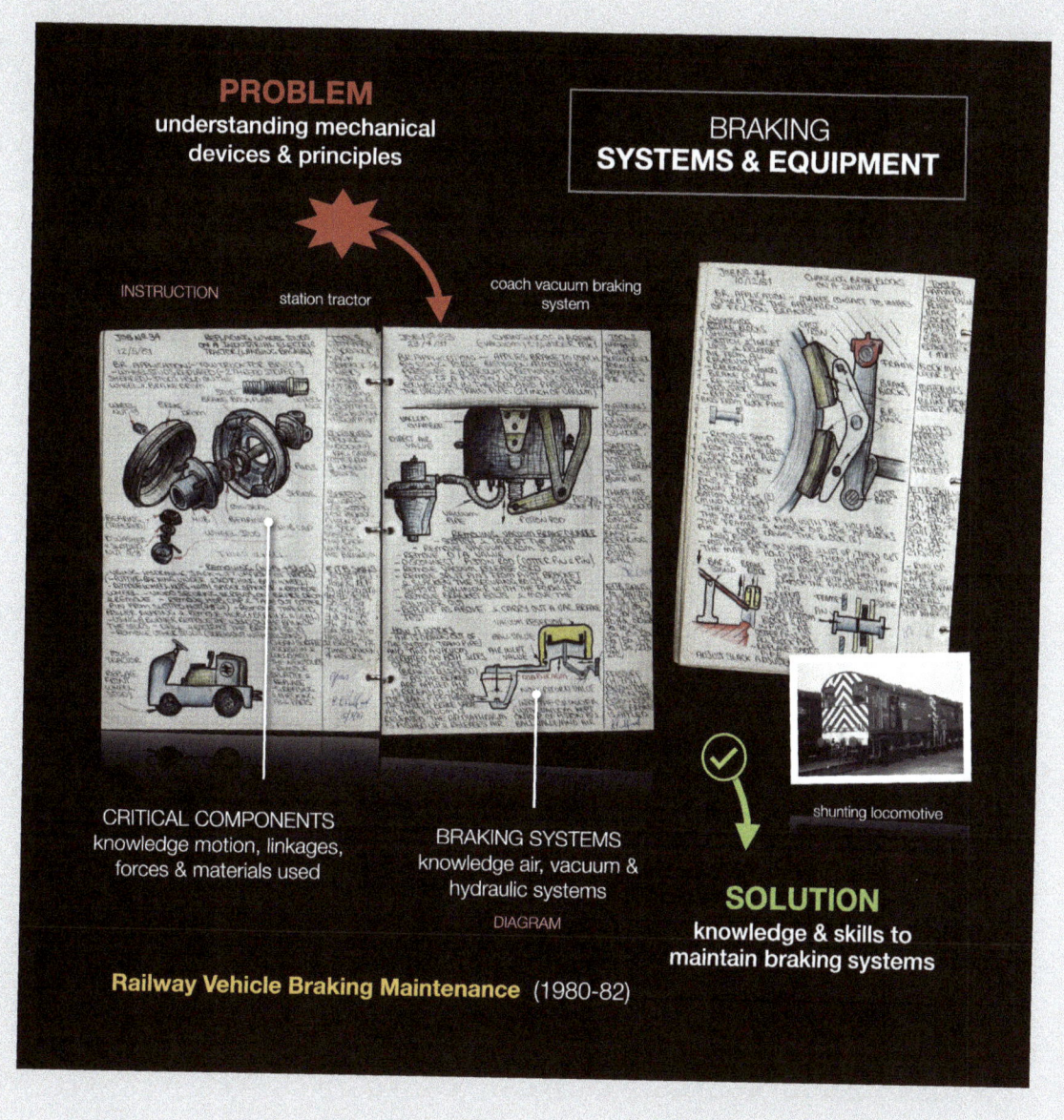

C4 MECHANICAL MAINTENANCE [1]

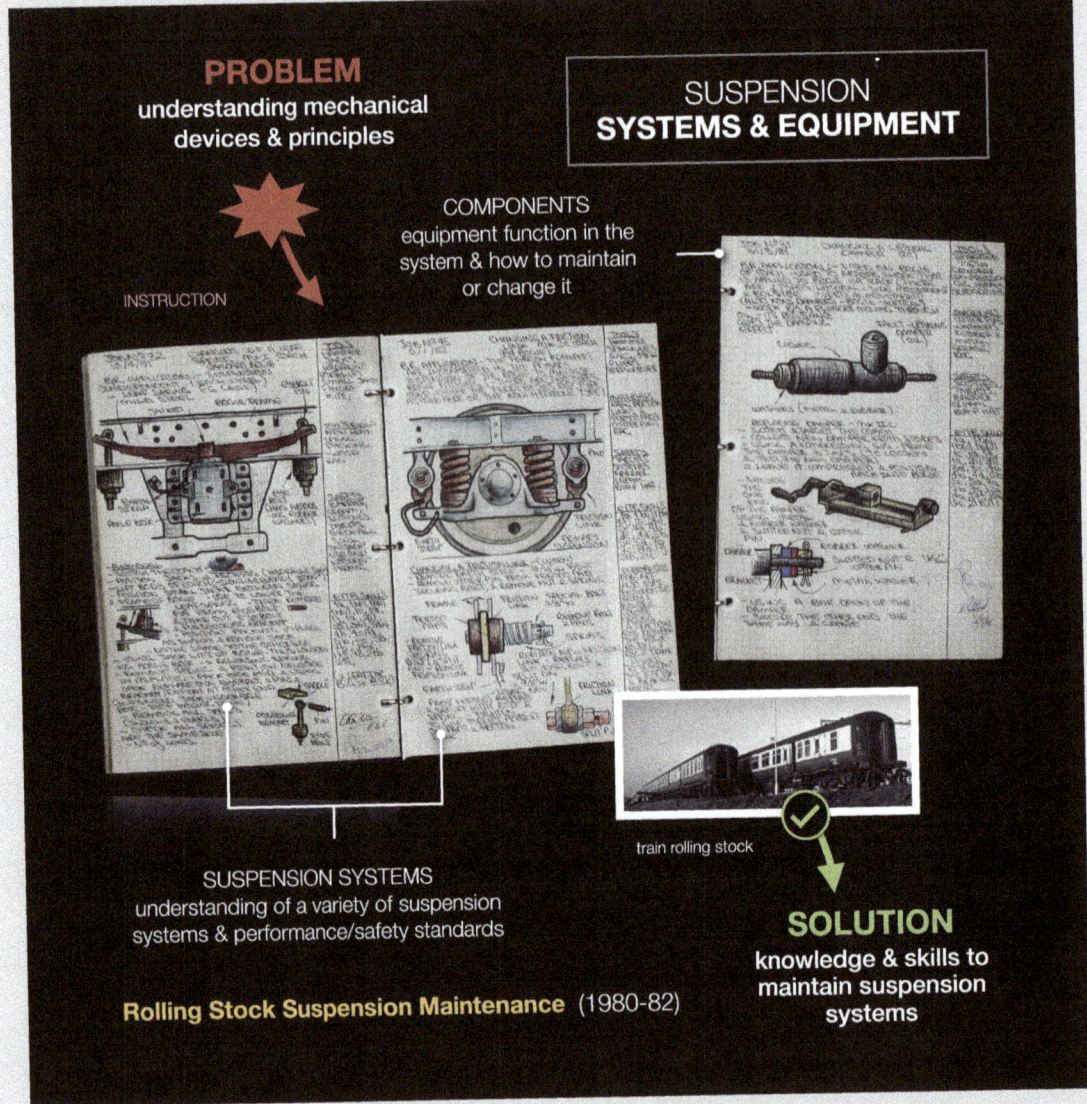

C5 MECHANICAL MAINTENANCE [2]

Mobile & Gantry Crane Routine Maintenance (1982)

C6 MECHANICAL MAINTENANCE [3]

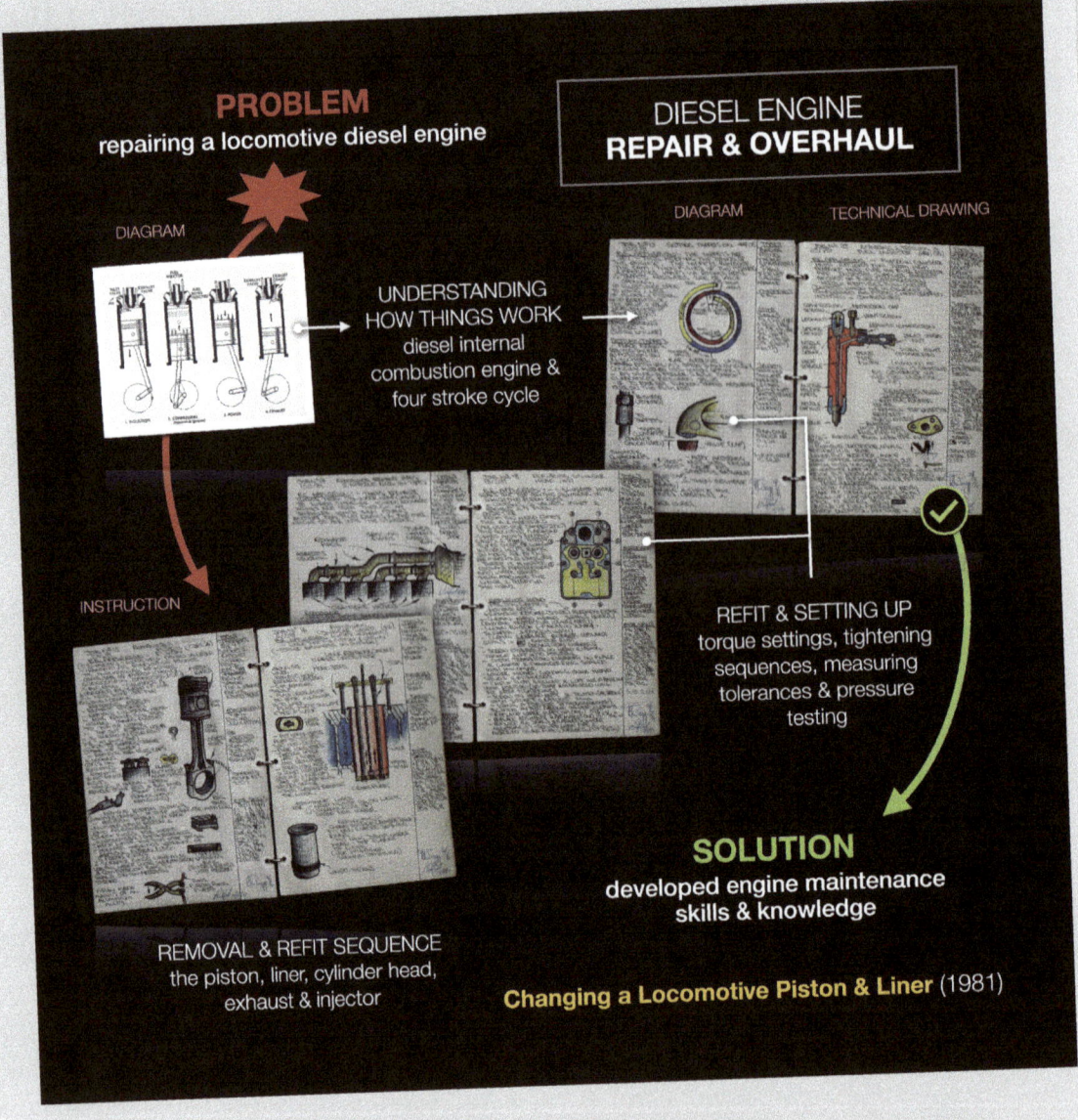

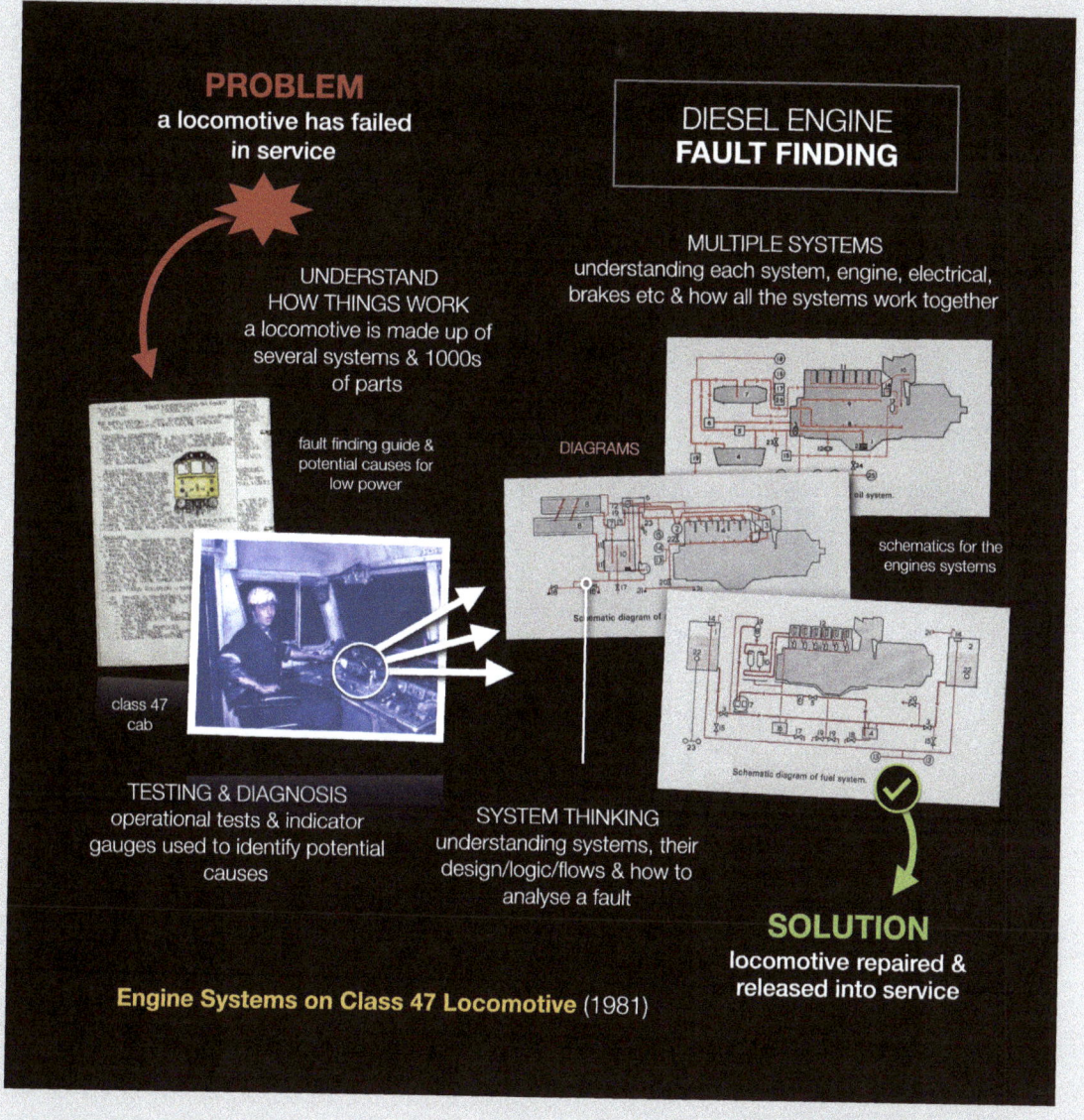

C8 FAULT FINDING - MECHANICAL [1]

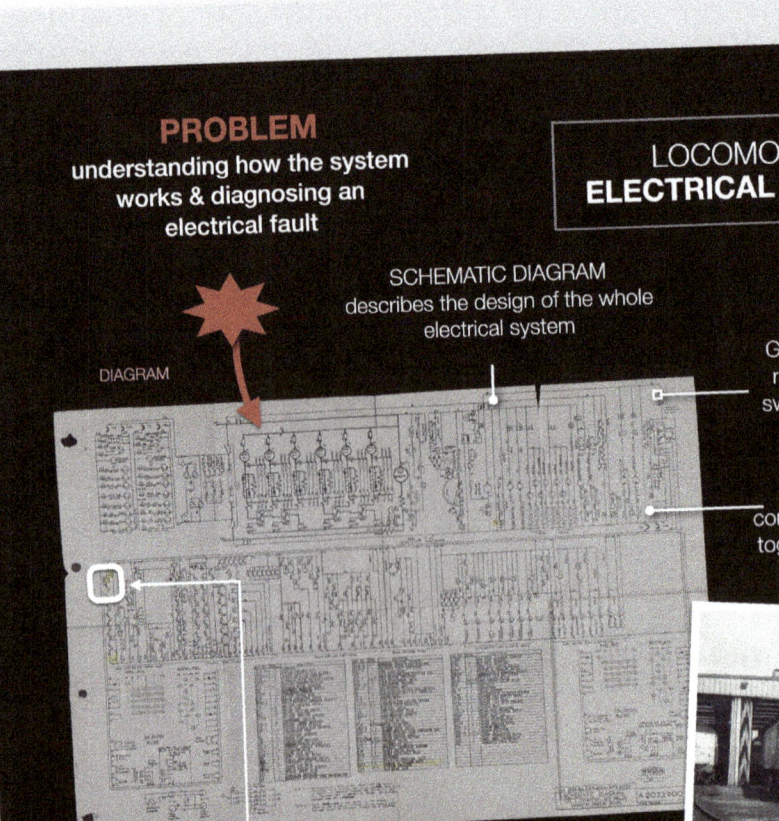

PROBLEM
understanding how the system works & diagnosing an electrical fault

LOCOMOTIVE ELECTRICAL SYSTEM

SCHEMATIC DIAGRAM
describes the design of the whole electrical system

DIAGRAM

GRAPHIC SYMBOLS
represent the part ie switch, motor, light etc

LINES
show how all the components are connected together & the circuit logic/design

class 47

FAULT FINDING
identify the specific circuit for that function then check components or the electrical supply to them

SOLUTION
fault diagnosed & rectified

Class 47 Locomotive Electrical Fault Finding (1981)

C9 FAULT FINDING - ELECTRICAL [2]

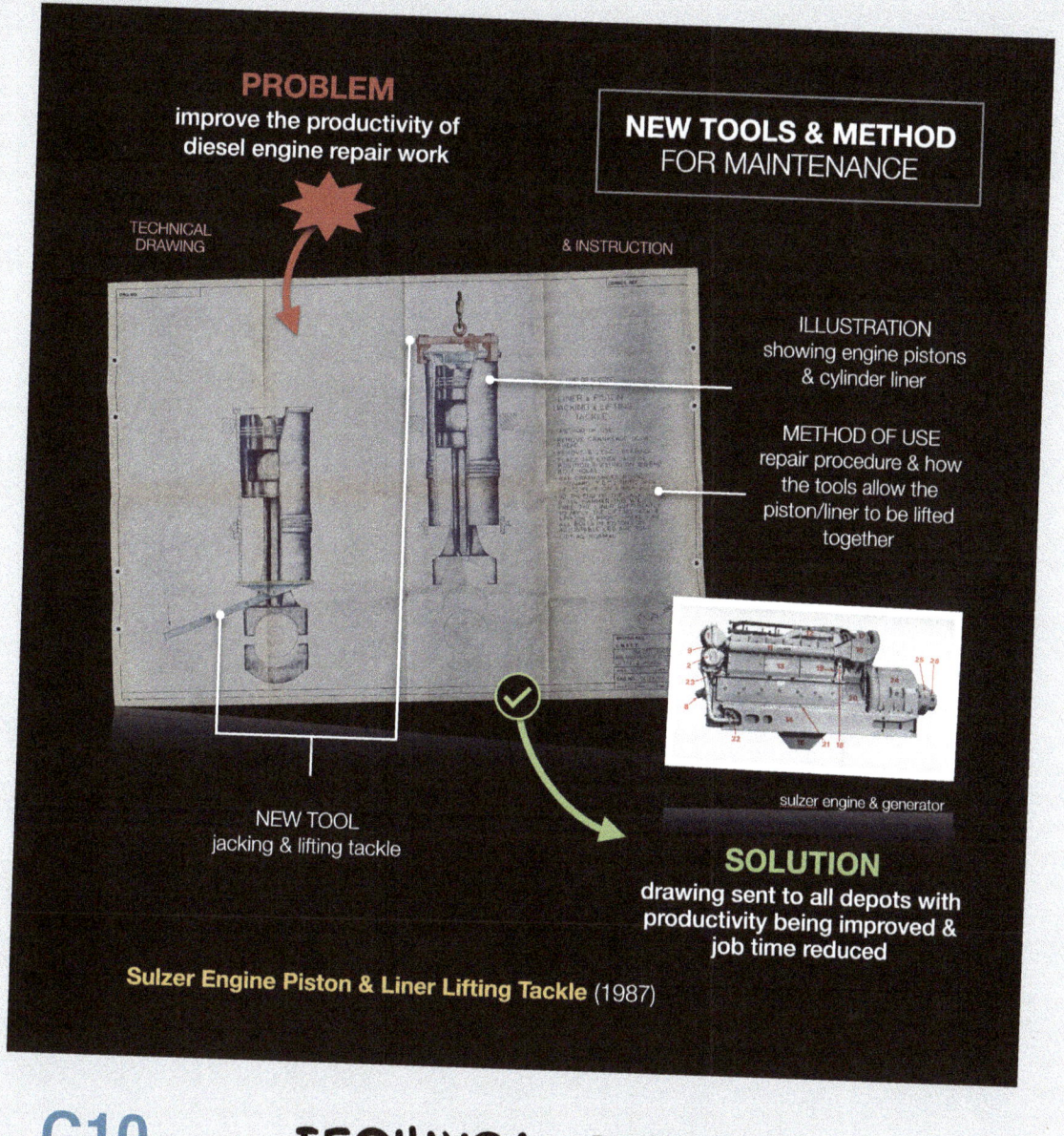

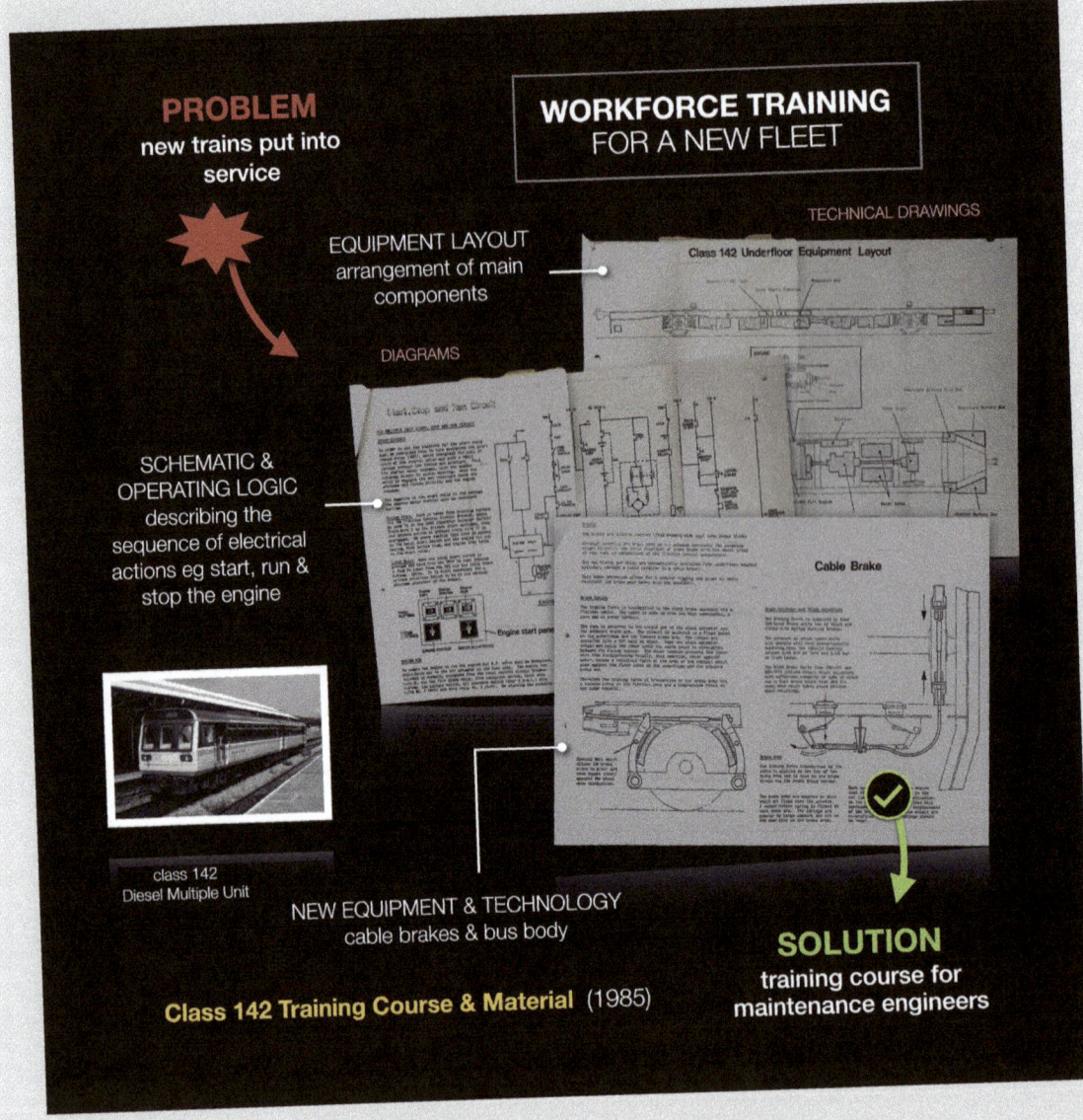

PROMOTING ENGINEERING & **APPRENTICESHIPS**

PROBLEM
recruiting apprentices & raising awareness in schools

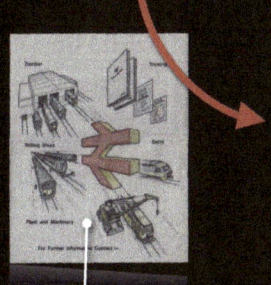

BROCHURE

MY DESIGN FOR A POSTER
the graphic design was based on my own apprenticeship & the places I worked

NATIONAL POSTER
thousands printed to provide resources for UK schools, career fairs & internal recruiters

ARTICLE
about me in 'railnews' British Rails national monthly newspaper

SOLUTION
national recruitment campaign run at career fairs & in schools

National Campaign Poster (1988)

C12   INSPIRING YOUNG PEOPLE

INSIGHTS on learning

USE REAL PROBLEMS AS LEARNING OPPORTUNITIES
Learning by doing real work, solving real problems & understanding the relevance of knowledge & skills

LEARNING FROM EXPERTS
Learning from others who have solved the problems before. Working in real teams develops social skills.

LOG YOUR LEARNING
Logging our achievements forces us to reflect on & consolidate what we have learned, this gives us a foundation to build on... and accelerates our learning

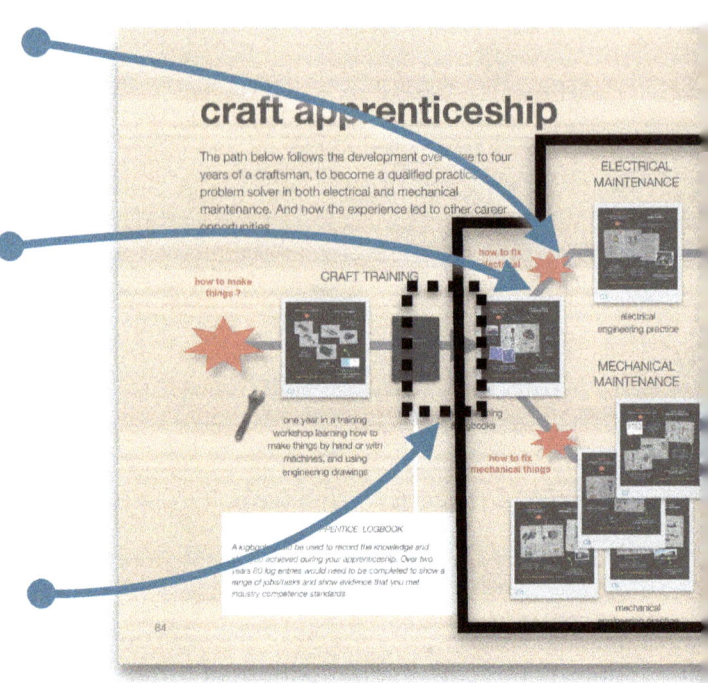

LEARNING and DEVELOPMENT TIPS

- ☑ learn by **doing real work** & solving **real problems**
- ☑ learn from **experienced & expert others**
- ☑ reflect on & **log** what you've learned or achieved

& practical **problem solving**

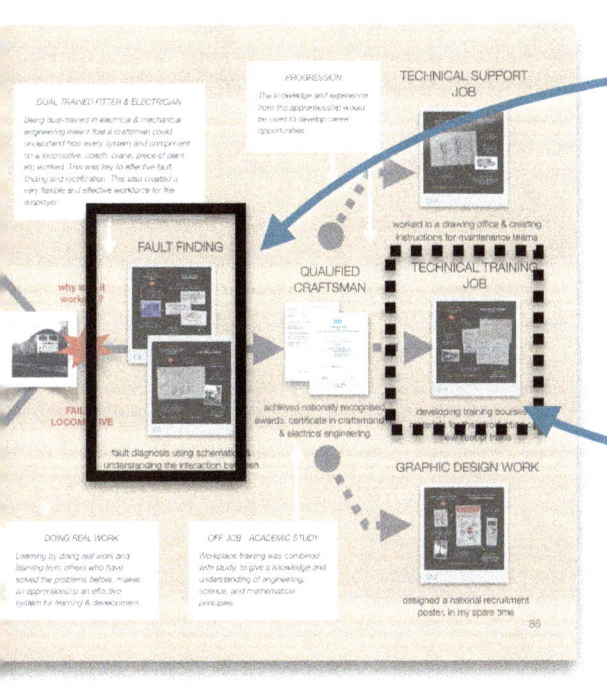

UNDERSTAND HOW THINGS WORK
engineers look at problems & things as 'systems' and use schematics to quickly understand why things work and then work out why it doesn't work

KNOW HOW TO DO THINGS
share & document knowledge describing the most effective & correct way to carry out a task

PROBLEM-SOLVING TIPS

- ☑ use systems thinking to understand **how things work**
- ☑ **identify the causes** using logic & reason
- ☑ using **best practice methods** to implement solutions

transferable skills

FIXING (& MAKING) THINGS

Engineers look at problems & things as 'systems' and use schematics to quickly understand how, and why things work and work out why it doesn't work, then use practical skills to solve the problem. The three skills below can used to help solve many problems :

use diagrams to quickly
UNDERSTAND HOW THINGS WORK
C3, C4, C7, C8, C9, C11

use diagrams to quickly
DIAGNOSE PROBLEMS
C9

use technical drawings to correctly
MAKE & REMAKE things
C1, C10

Engineering teaches us to **look at problems differently**, to see beyond the details, and to understand how things work. This 'systems thinking' approach can help us solve non-technical problems ….

DESIGNER
draughtsman

the job of a **designer**

DESIGNING THINGS

As a designer, I would be part of a team that would use drawings to develop the form, look, and workings of trains. We would then produce technical drawings to enable manufacturers to build the vehicles. Tasks included :

feasibility studies &
undertaking engineering **design** to develop
new products

produce drawings, diagrams & schematics to allow
manufacturers to **build** things

investigating problems with existing products
(on trains being **maintained**) & improve

Designers deal with new problems. This can result in improvements or modifications to existing products or the development of new products.

New problems require the designer to have the theoretical knowledge to solve problems in developed technologies and using proven analytical techniques.

They work from a design brief and develop ideas using concept sketches, then produce technical drawings to accurately communicate with any manufacturer (anywhere in the world) and get things made.

job profile

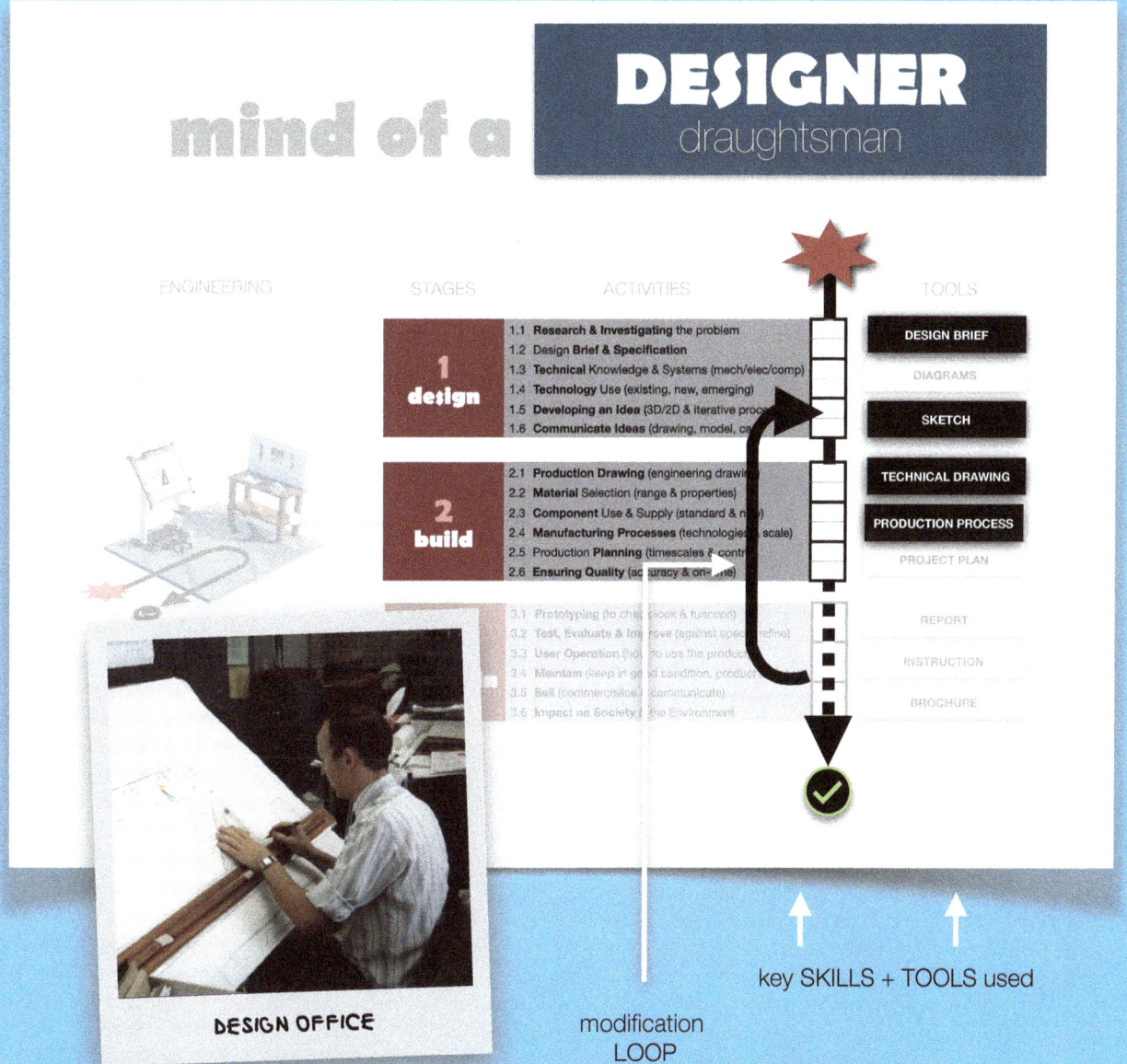

inside the **design office**

THE DESIGN OFFICE

British Rail designed and built its trains. The design offices of the Railway Technical Centre were responsible for designing and developing new fleets and improving existing vehicles.

DESIGNER/DRAUGHTSMAN

Before I entered the Design Office I would spend time in the engineering works to understand **how things are manufactured** and to know what was possible and capable of being produced.

Then in the drawing office training school learned how to produce **technical drawings** to the required standard. These would be part of a formalise system with a series of drawings (general arrangement, assembly & part details) that would be checked and endorsed by senior engineers.

A broad range of **traction & rolling stock projects** would be worked on from initial ideas that required concept sketches or illustrations, to full production of new vehicles or systems. These designs would often backed up with calculations.

Computer-aided design (**CAD**) was introduced in the mid-1980s and changed the approach to design and the skills needed. It allowed things to be drawn full size, created in 3D, seen from multiple views, and quickly duplicated/modified.

workplace

INTERCITY

The Intercity Business provided centre to centre travel across the UK and operated high speed trains.

In 1982 the British Rail regions were abolished and replaced by business sectors, Intercity was one of three passenger sectors along with Regional Railways and Network SouthEast.

RAILWAY TECHNICAL CENTRE

The Railway Technical Centre, Derby. was established in 1964 to concentrate British Rails engineering expertise. British Rail described it as the largest railway research complex in the world

Railway Technical Centre

APT & Intercity Design Office, Derby
The Technical Headquarters of the British Railways Board and the Department for Mechanical & Electrical Engineering.

The Design Office would design New Fleets of trains and develop, improve the Current Fleets.

The Projects included :
- New Coach : Mark 3B
- New Intercity Fleet : IC225 Concepts
- Electric Locomotive : Push -Pull Operation
- High Speed Train : Rheostatic Brakes
- APT : Modifications
- CAD : User Guide & Development Projects

Railway Technical Centre, Derby

APT & INTERCITY DESIGN TEAM

The advanced passenger train (APT) & Intercity Design Office had a team of 20-30 engineers working on rail vehicle mechanical, electrical & interior designs. They were supported by other specialist departments ie vehicle dynamics, brakes etc

QUALIFICATION

Higher National Certificate (HNC) qualification in electrical & electronic engineering

Intercity Design Team

Design Team
The design team was made up of :
- Mechanical Engineers
- Electrical Engineers
- Interior Designers

Designer Draughtsman Role
- Feasibility studies, analyses and design calculations of a stress, dynamics
- Production of drawing, diagrams and schematics.
- Investigation into problems
- Production of reports, technical modifications and maintenance instructions

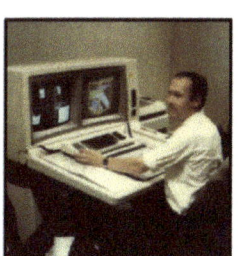

Producing Engineering Drawings & Technical Illustrations

design **technical** knowledge

PASSENGER TRAINS

There are two basic types of passenger train, those consisting of a locomotive pulling or pushing unpowered passenger coaches and self-propelled trains (multiple units).

The size and formation of a train is a business decision and vehicles with different features will often be required. Coach design variations often include first class, second class, catering or buffet cars, sleeping cars, and the addition of train crew accommodation.

COACH DESIGN

Passenger coaches are designed to safely carry several customers, in comfort and at speed. Specific design and maintenance choices will also need to consider the 30-50-year lifespan of a rail vehicle. Some of the technical challenges include :

- Vehicle Structure (including crashworthiness)
- Bogies (ride stability & quality)
- Seating (interior design & passenger capacity)
- Lighting, Heating & Air Conditioning (systems)
- Toilets, Luggage, Disability & Walkways
- Customer Information (systems, labels, etc)
- Safety : Boarding, Alighting & Emergencies
- Systems (Braking, Air Con, Electrical etc)
- Inter-vehicle connections

Vehicles are linked to one another or other vehicles with couplings and have inter-vehicle cables, air pipes, and often passenger gangways.

intercity trains

KNOWLEDGE OF
ENGINEERING **DESIGN** & **MANUFACTURING**

PASSENGER ENVIRONMENT

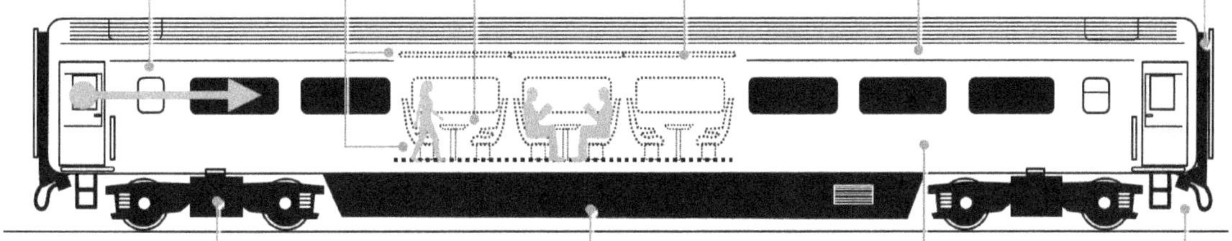

TOILET · LUGGAGE SPACE · SEATING LAYOUT · LIGHTING · AIR CONDITIONING · GANGWAY

BOGIES: WHEEL-SETS, BRAKES & SUSPENSION · BRAKING, AIR CONDITIONING, BATTERY & POWER EQUIPMENT · VEHICLE STRUCTURE · INTER VEHICLE CONNECTIONS

MECHANICAL & ELECTRICAL SYSTEMS

BRITISH RAIL MARK 3 COACH

MAXIMUM SPEED	125mph
LENGTH	23m

DESIGN VARIATIONS

- ☐ FIRST & SECOND CLASS SALOONS
- ☐ CATERING & BUFFET CARS
- ☐ SLEEPING CARS
- ☐ TRAILER CARS (FOR HST SETS)

Title: **BRITISH RAIL MARK 3 COACH**

ALL EXISTING & NEW INTERCITY **TRACTION & ROLLING STOCK**
including high speed trains, locomotives, rolling stock & new train concepts

design projects

The path below shows my development as a technician trainee and designer, and some of the projects undertaken over three years, including the early introduction of computer-aided design (CAD).

TECHNICIAN TRAINING PROGRAMME

DESIGNER DRAUGHTMAN JOB

DESIGN & TECHNICAL KNOWLEDGE

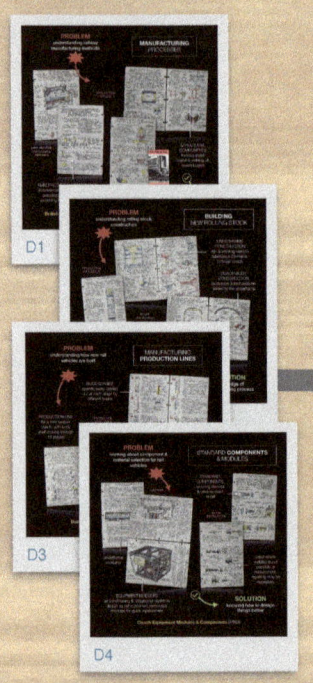

spent 6 month in Railway Engineering Workshops, understanding the manufacturing processes, techniques, tools & materials used

learning to produce technical drawings to an industry standard & document designs

completed an HNC in Electrical & Electronic Engineering & awarded technician training certificate

I joined the APT Design Team in 1983, which became the Intercity Design Team in 1984

learning about specific railway systems & industry design standards

BRITISH RAIL ENGINEERING WORKS

Up until the mid-1980s British Rail designed its trains and built them in its workshops, British Rail Engineering Limited.

ADVANCED PASSENGER TRAIN : DESIGN OFFICE

APT was a tilting high speed train, service prototypes ran from 1980-86. Plans for a production version were abandoned in 1984. In spite of its troubled history the design was highly influential and directly inspired other trains, including the Intercity 225.

improving existing traction

TRACTION : PROJECTS

CAD ENGINEER JOB

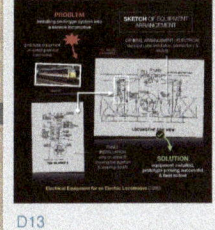
D13
design for a new system on an electric locomotive

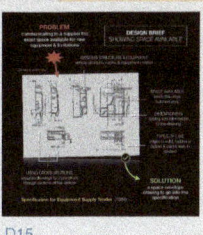
D14 D15
design specification for a new braking system on a high speed train

D16
I was seconded to the CAD team to produce a user guide, to deliver training & develop new electrical design package

specification for new trains

NEW TRAIN : PROJECT

the concept of the InterCity 225 had been developed from the best features of the APT & HST

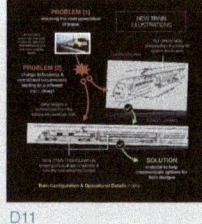
D11 D12
business requirement & specification for a new fleet of trains

communicate the technical challenges & some design details

intercity 225 with a locomotive & nine coaches, in service on the newly electrified East Coast Main Line

new rolling stock

ROLLING STOCK : PROJECTS

D9
interior lighting installation for a new coach

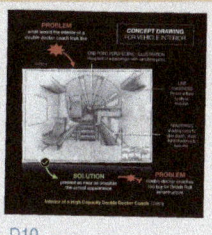
D10
feasibility study for of double decker coaches

NEW TRAINS : PROCURED

In the mid-1980s, British Rail moved away from designing and building its trains. New trains would be specified and procured through competitive tendering. Today UK trains are designed and built by lots of different suppliers & the fleets are owned by private companies.

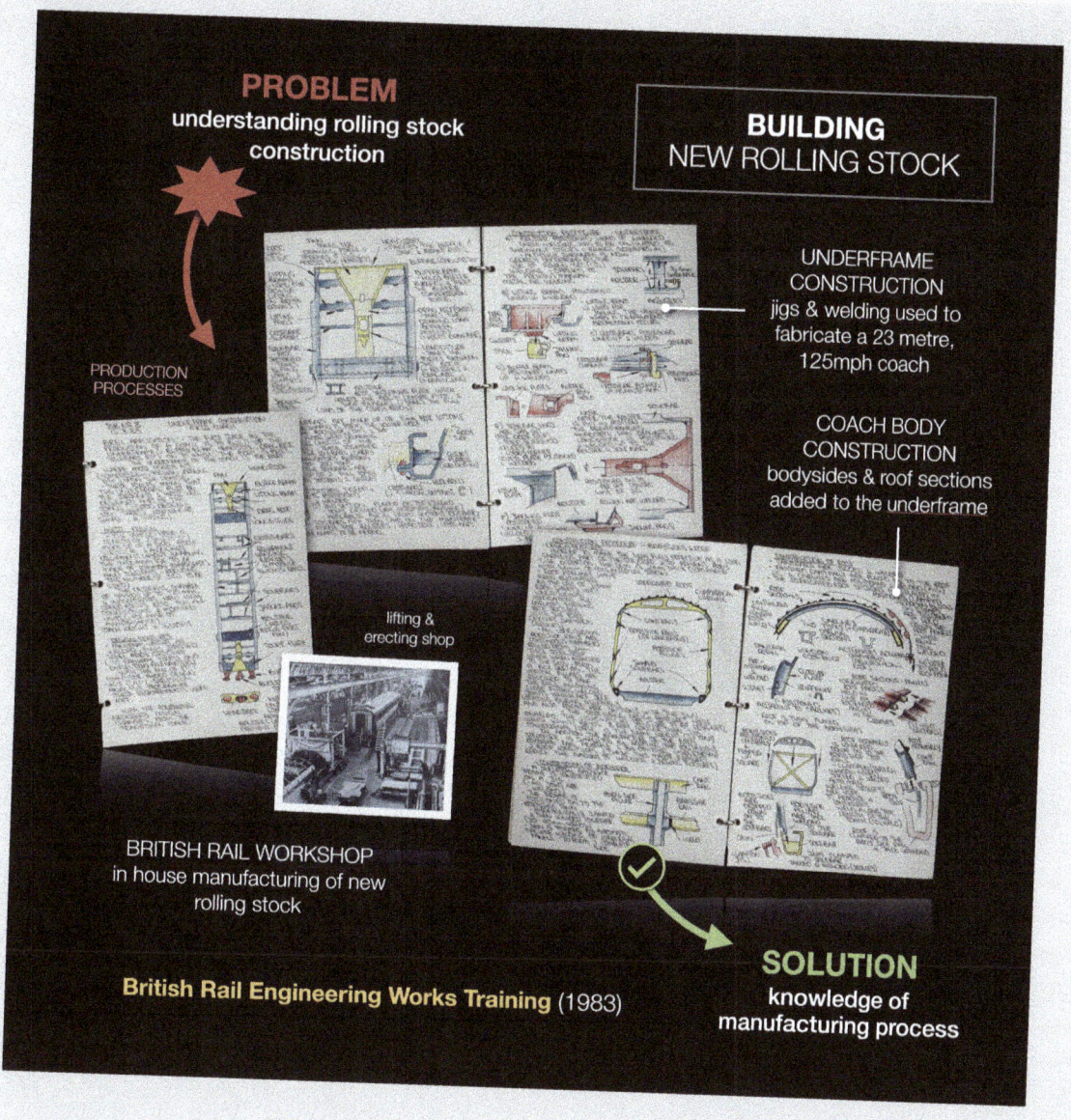

D2 — MANUFACTURING [2]

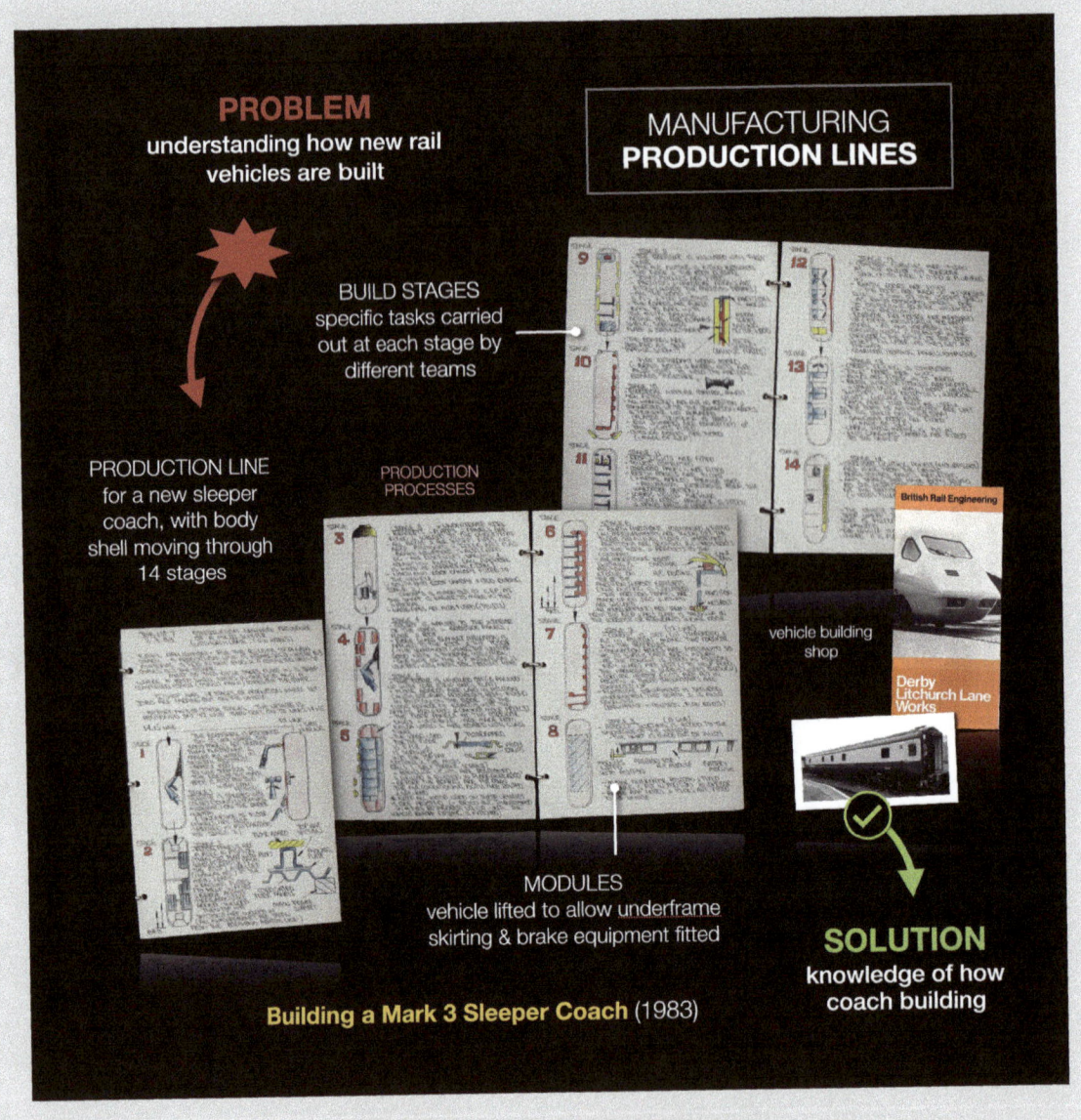

MANUFACTURING [3]

D3

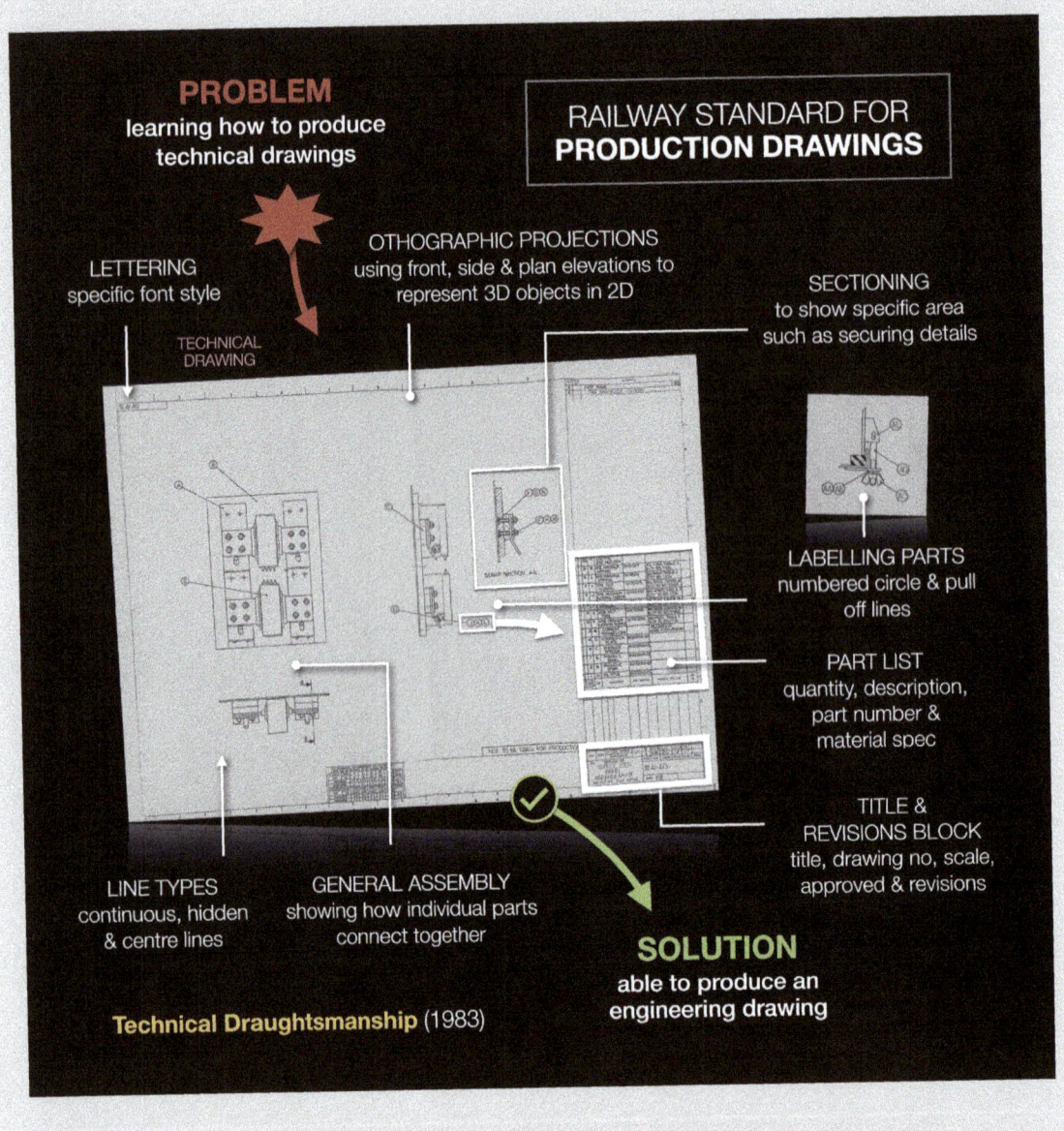

D5 — DRAWING OFFICE TRAINING

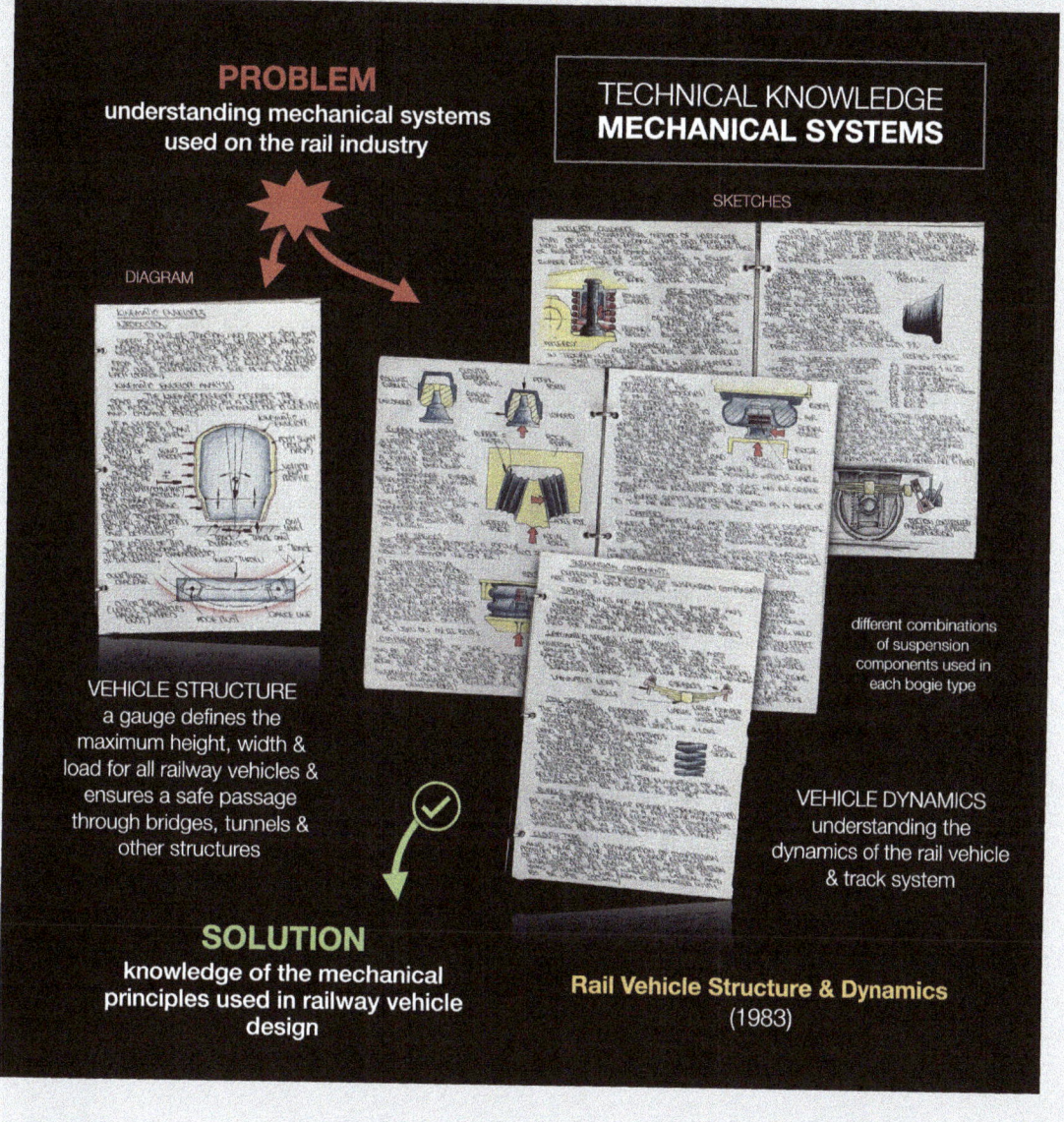

D6 — MECHANICAL SYSTEMS

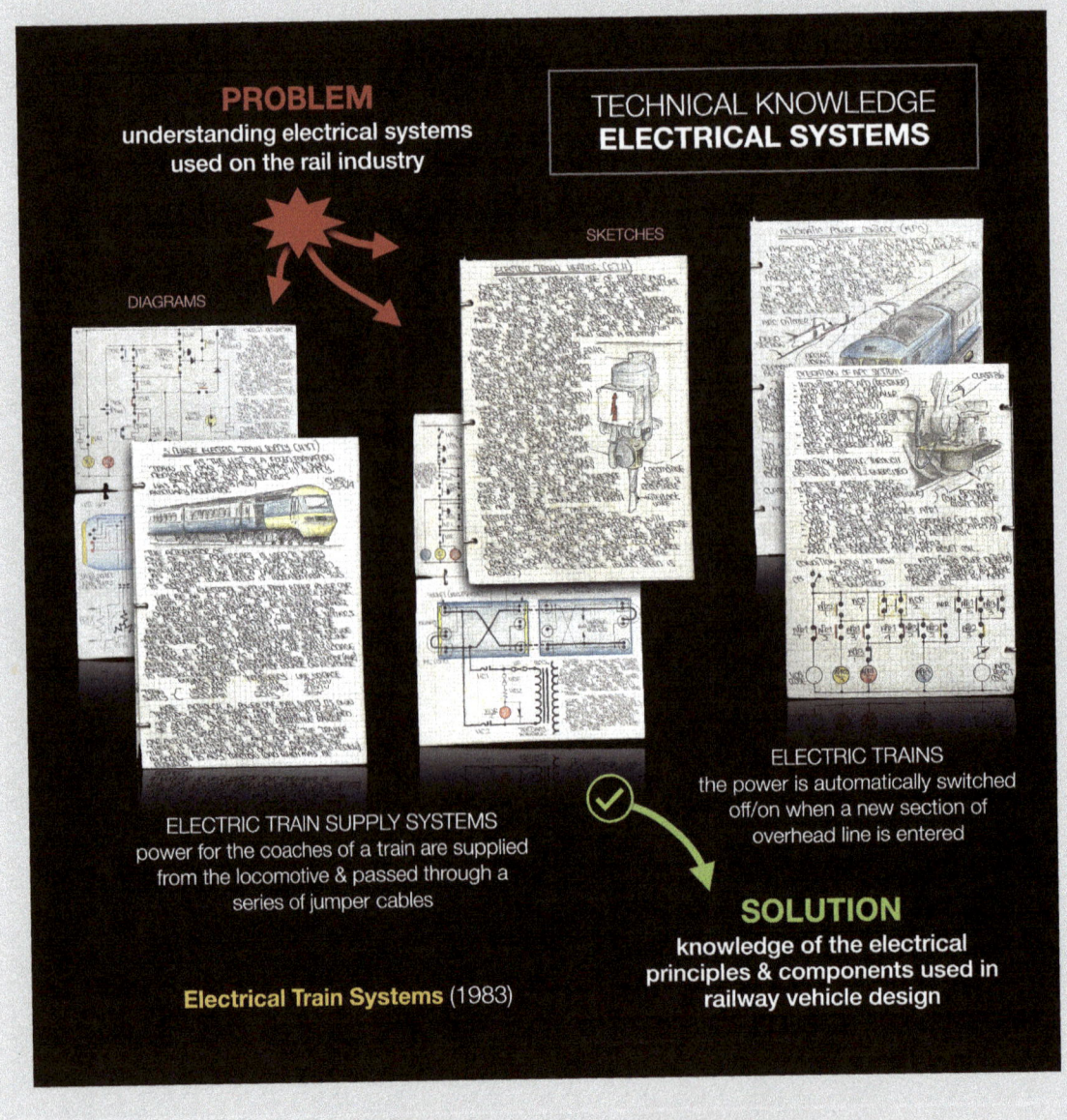

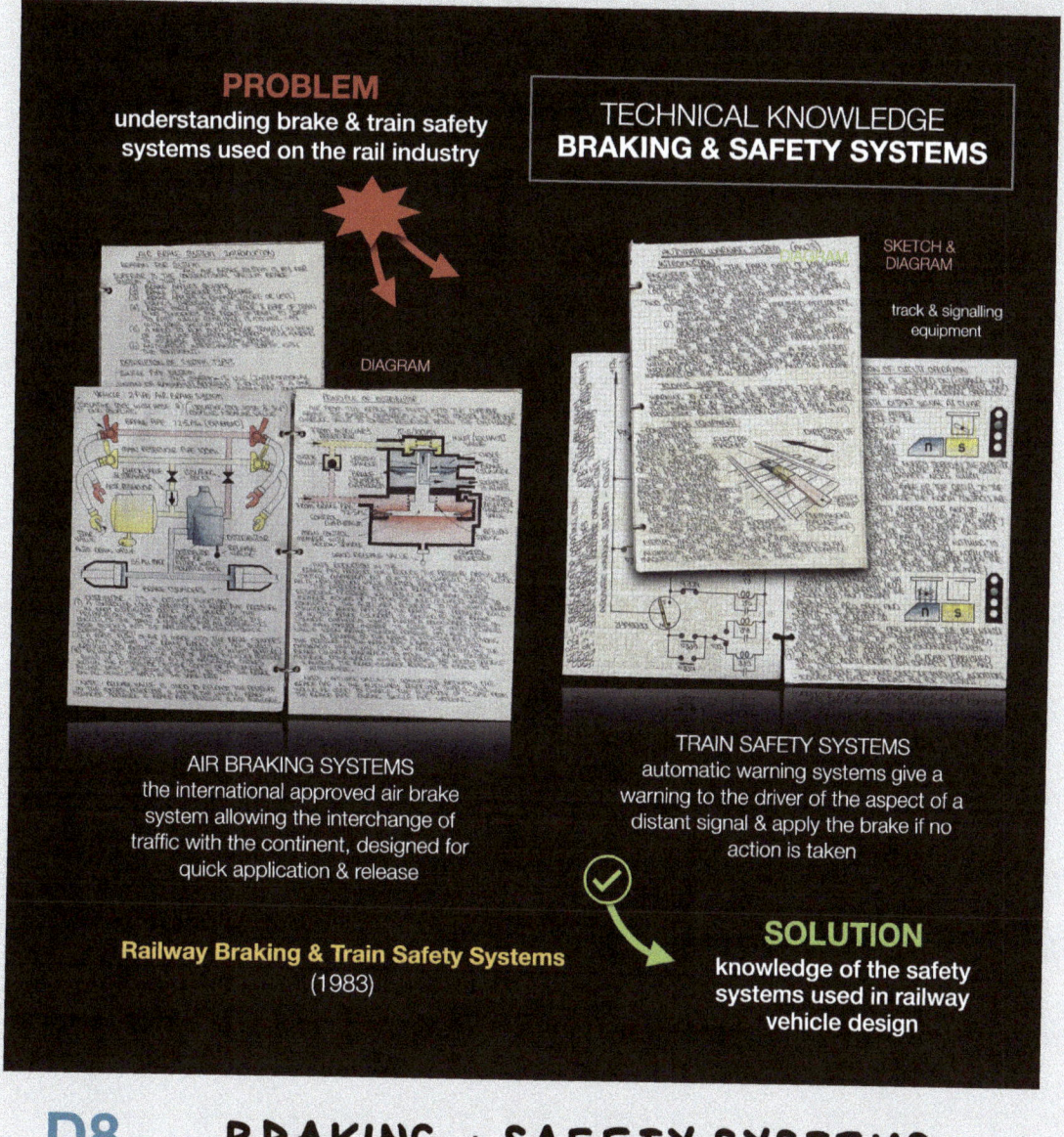

D8 BRAKING + SAFETY SYSTEMS

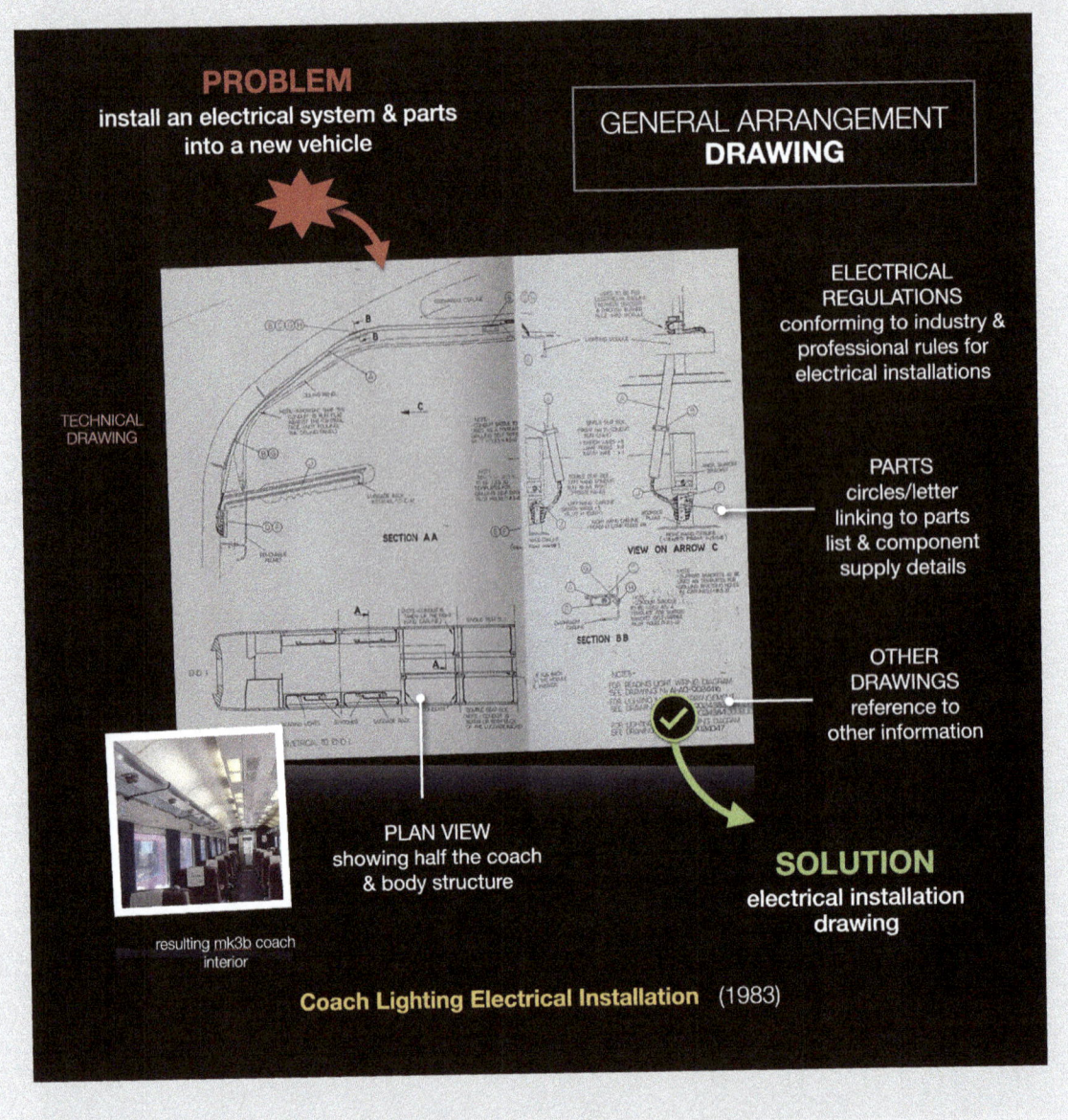

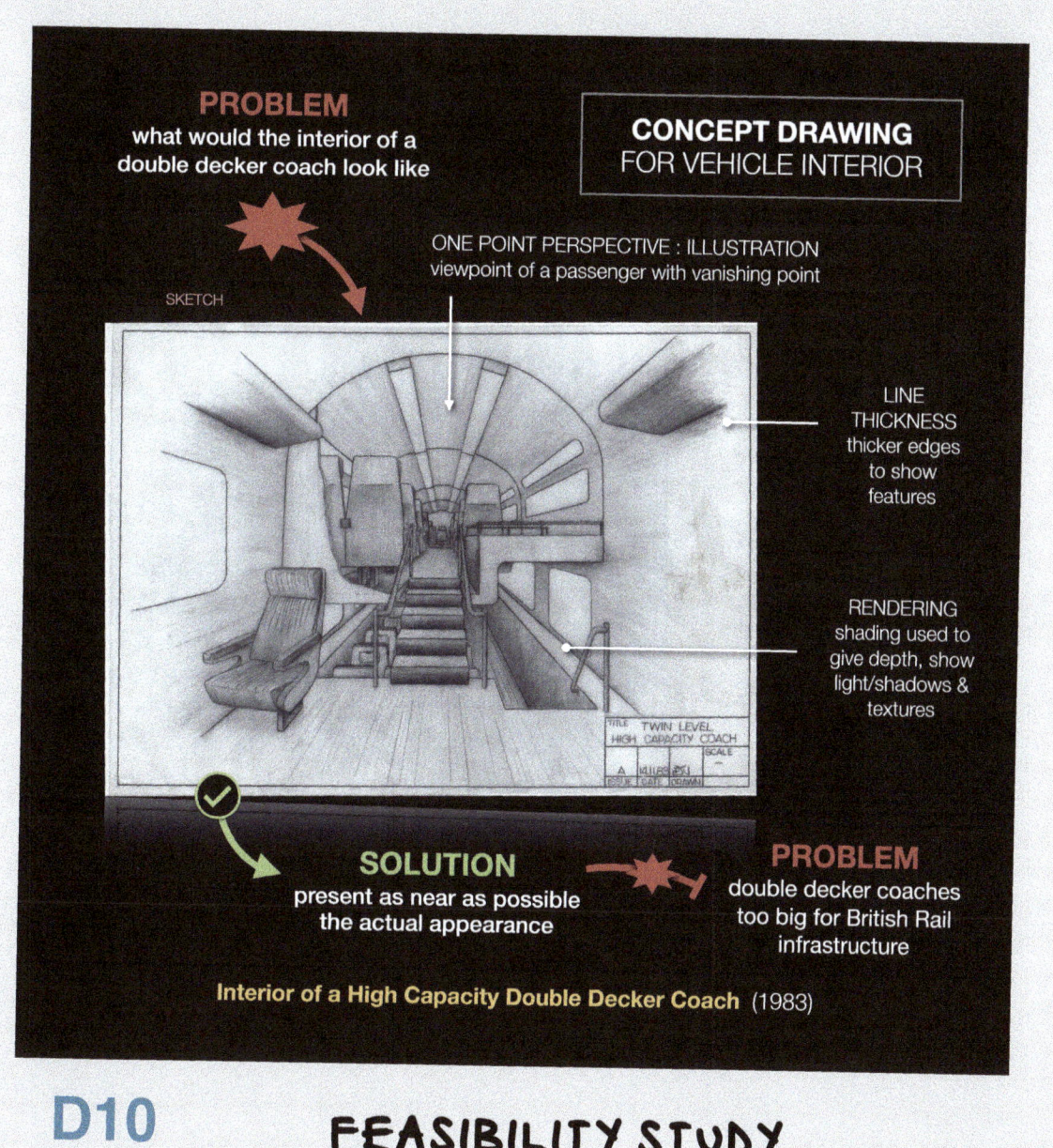

Interior of a High Capacity Double Decker Coach (1983)

D10 FEASIBILITY STUDY

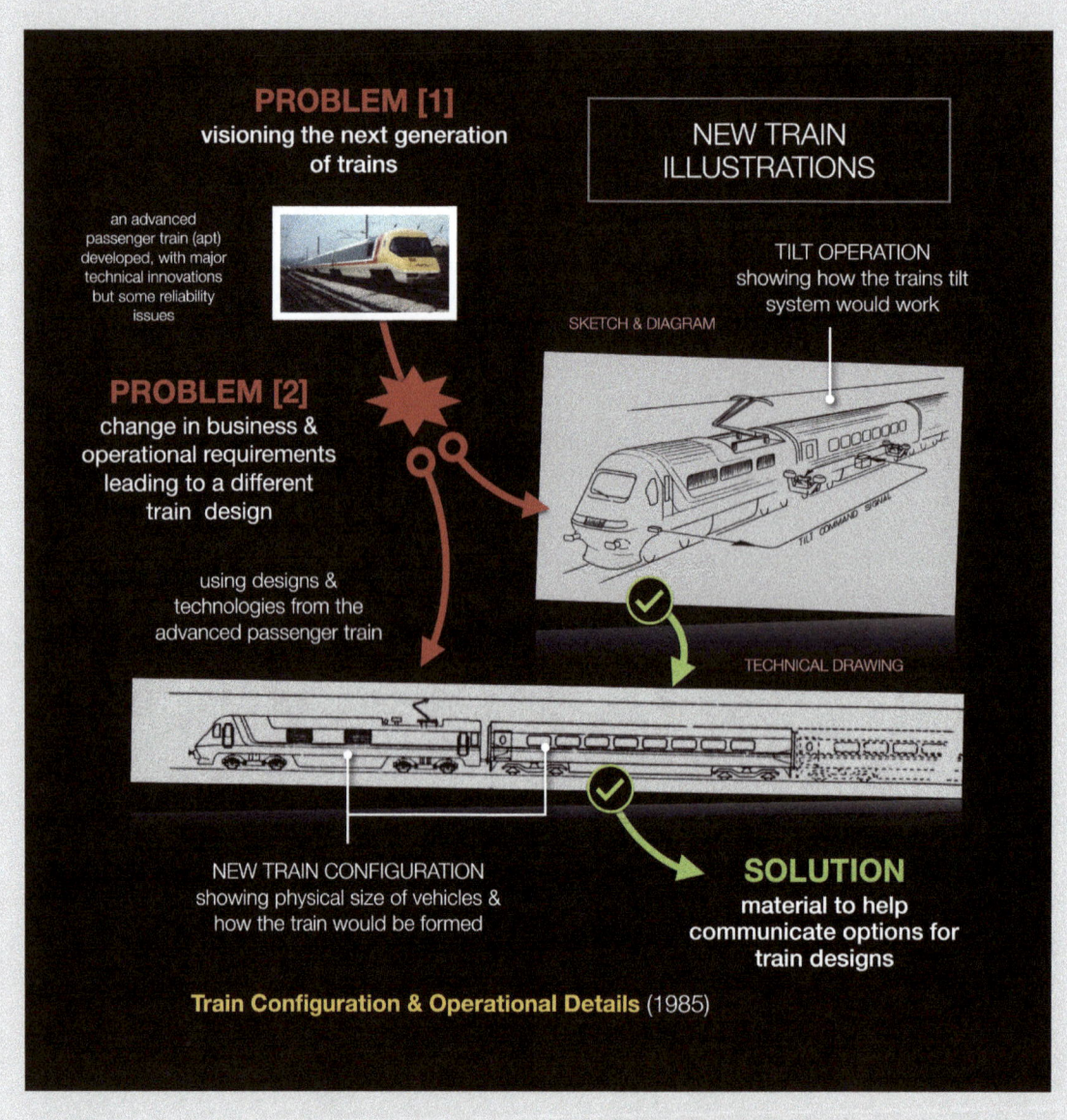

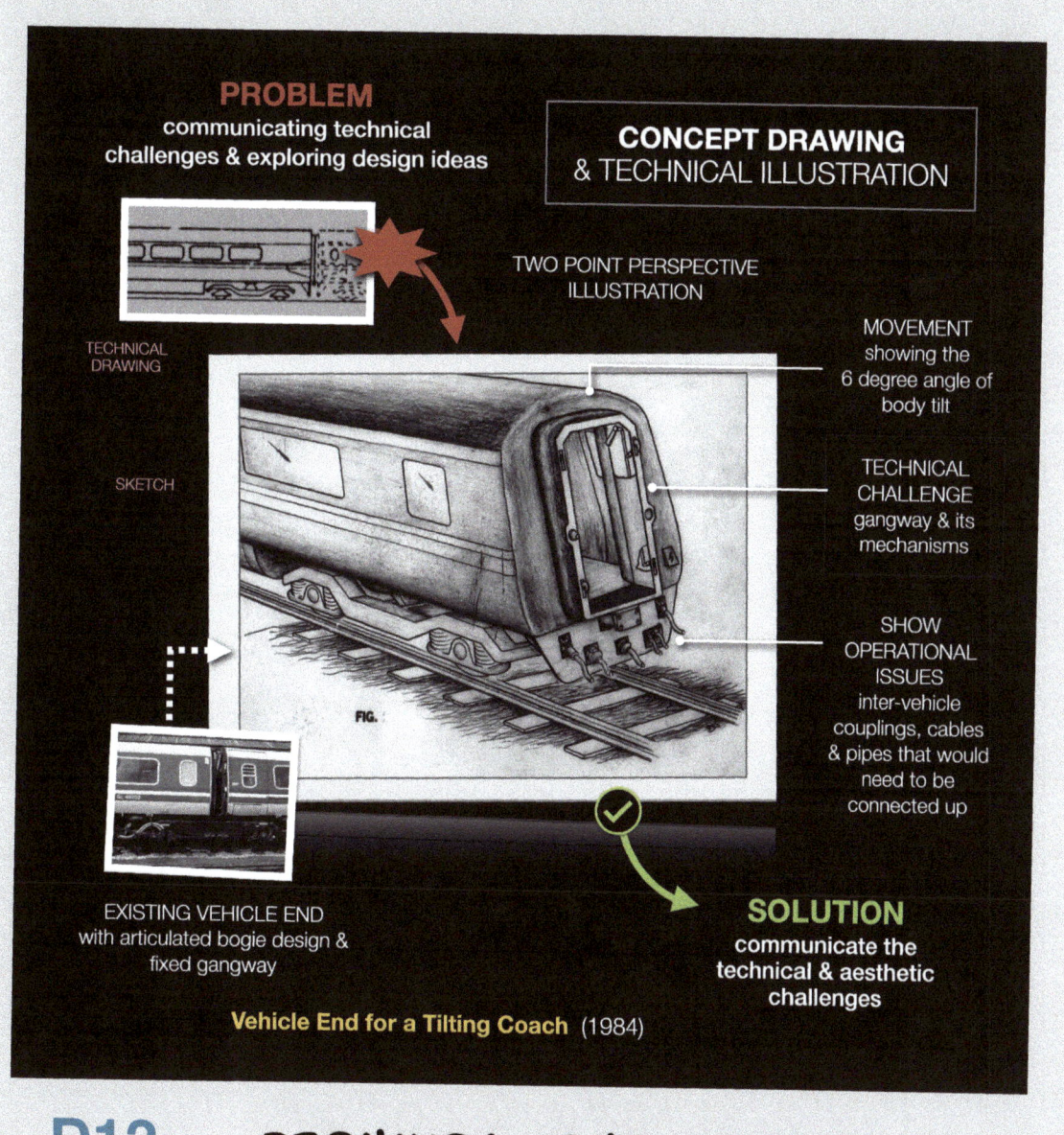

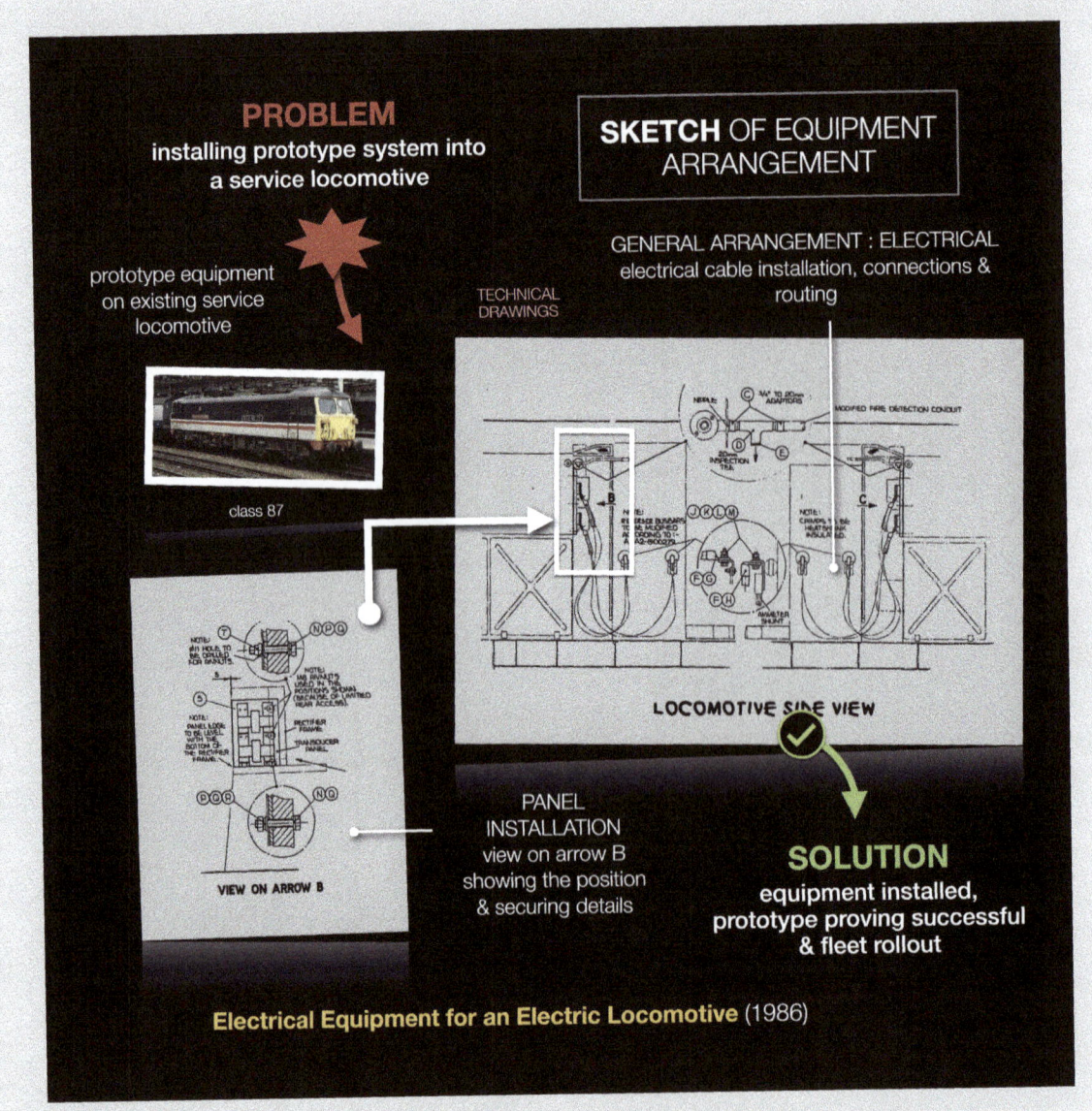

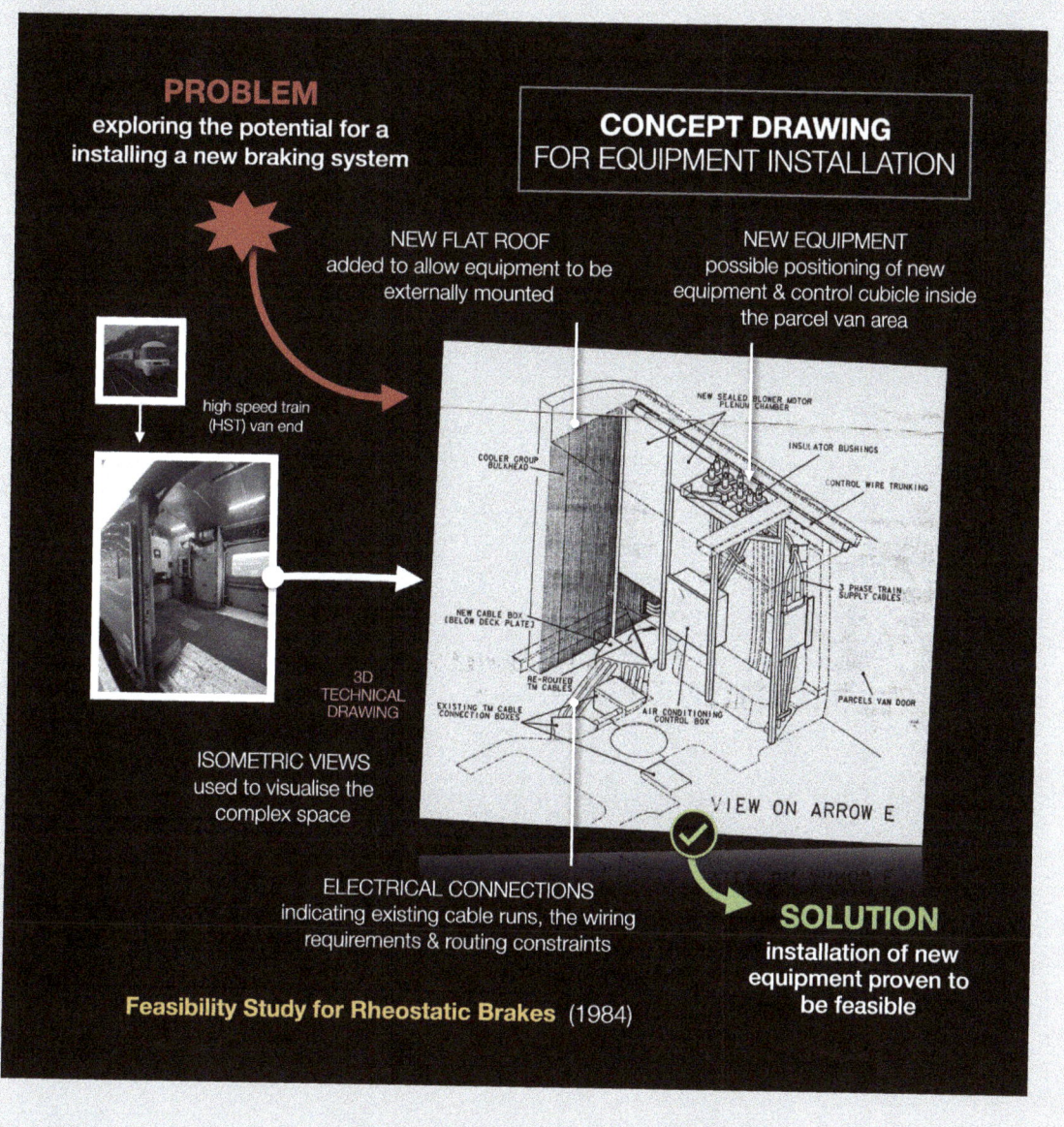

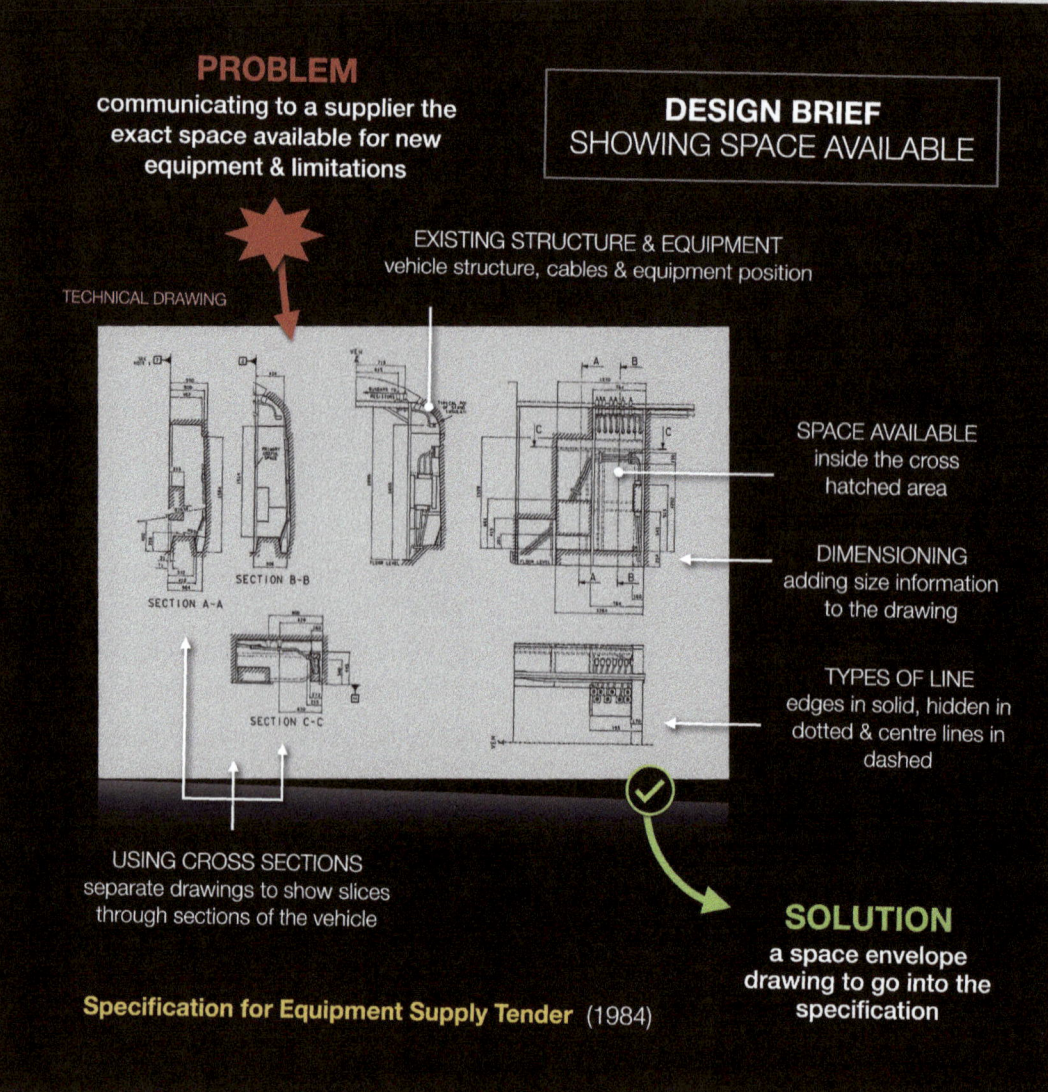

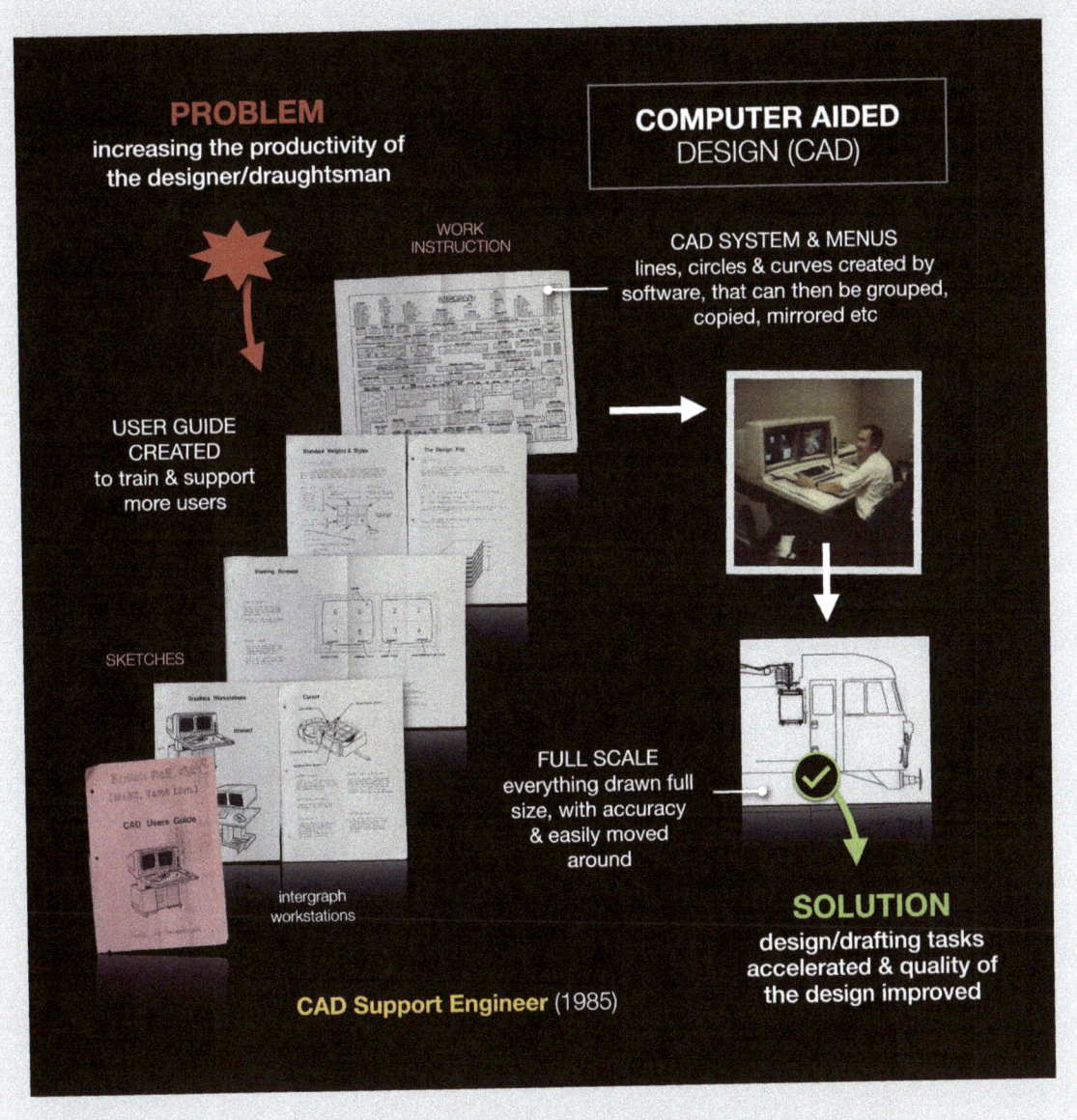

D16 COMPUTER AIDED DESIGN

INSIGHTS on manufacturing

APPRECIATION OF MANUFACTURING METHODS

Understanding how things are made is important to design, you need to know what is possible and capable of being produced.

TECHNICAL DRAWINGS TO COMMUNICATE WITH MAKERS

Technical Drawing enables quick & accurate communication with any maker, about how something can be made & a record of development

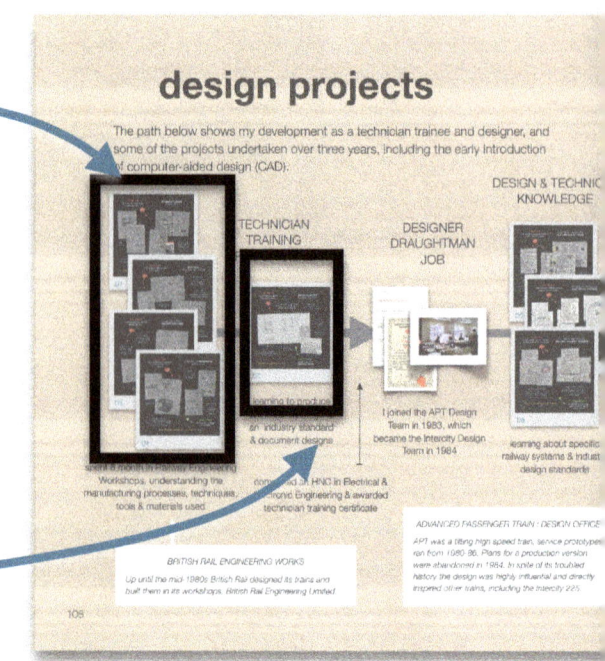

BENEFITS of MANUFACTURING awareness :

- ☑ understand **how things are made** & the constraints
- ☑ know why systems & things have been designed
- ☑ using technical drawings to **communicate with makers**

& engineering **design**

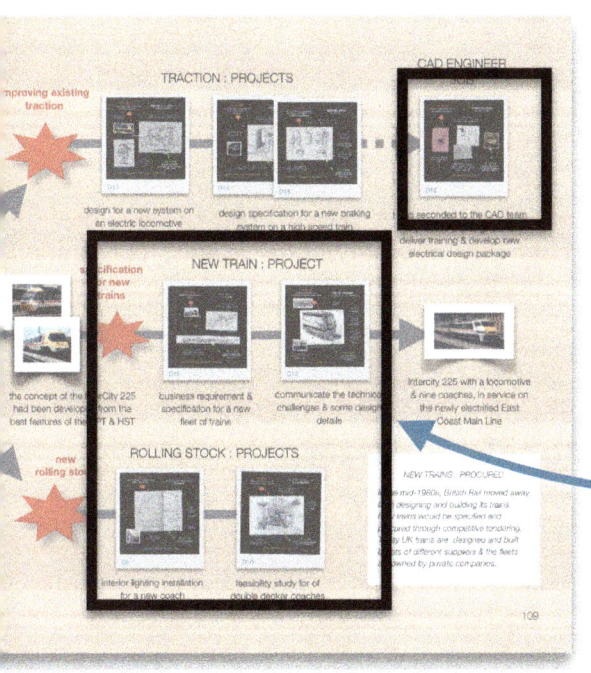

COMPUTER-AIDED DESIGN (CAD)
The computer screen replaced the drawing board allowing designs to be created full size, common elements to be stored in libraries, alterations to be made quicker and 3D models to be built

USE SKETCHES TO CAPTURE + EXPLORE IDEAS
Sketching, concept drawings, and Illustrations help to capture & communicate ideas

---

SECRETS of DESIGN & drawings :

- ☑ drawings visualise our thinking & can **bring ideas alive**
- ☑ drawings show what something or the future **will look like**
- ☑ allow ideas to be discussed, developed, or dismissed

transferable skills

DESIGNING THINGS

Drawing can be used to capture, explore, and communicate any idea. These techniques can be used in other jobs to visualise thinking, present ideas and bring visions alive. The three skills below can used to help turn ideas into reality :

use sketches to quickly
EXPLORE & BRING ALIVE IDEAS
D9, D10, D11 & D12

use diagrams to quickly
UNDERSTAND MANUFACTURING
D1, D2, D3 & D4

technical drawings to
COMMUNICATE WITH MAKERS
D5, D6, D7 & D8

The discipline of technical drawing and the need to communicate lots of information to a maker or solution builder would help me create practical solutions to non-engineering problems.

RESEARCH
scientist

the job of a **research scientist**

INVENTING THINGS

My job was to develop solutions to complex engineering problems using existing or new technologies, through innovation, creativity, and technical analysis. Carrying out the following tasks :

investigate unusual or challenging train service & **maintenance** problems

invent, **design** & develop new innovative technical solutions or develop new theories

build prototypes

test (use & **maintain**) the prototype to evaluate effectiveness & commercialise the product

The problems are **unusual, challenging, and complex,** so require investigation to understand what is happening.

Once the problem is understood they then **innovate**, to create new solutions and develop them through experimentation, failure, and iteration.

Creating a technical or business vision of the future is exciting but difficult and risky, and requires us to apply a rigorous approach to design/build/maintain.

The innovative nature of the work meant that the design/build/maintain stages were repeated several times as the solutions moved from prototypes into products.

job profile

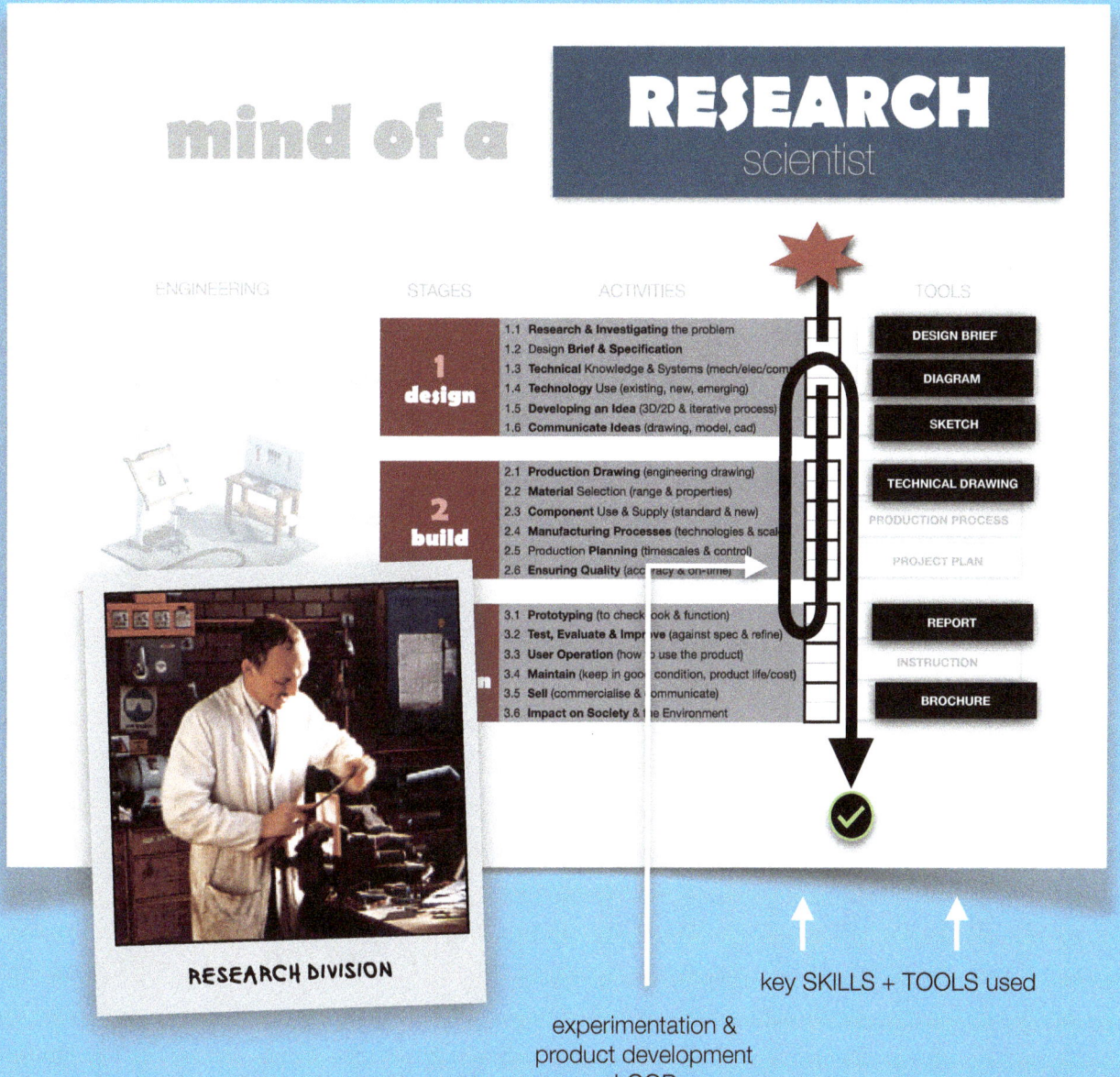

inside the R&D lab

RESEARCH & DEVELOPMENT LAB

The British Rail Research Division was established to improve railway reliability and efficiency while reducing costs and improving revenue. The electrification team worked on leading-edge projects to support vehicle and infrastructure aspects of electrification. Creating a technical and business vision of the future, working on a 10-15 year timescale. Projects would turn theory into operational hardware, moving from research into development and eventual implementation.

RESEARCH SCIENTIST

Exploring the technical feasibility and business potential of new concepts. I will show you how some of these innovative projects (pantograph, overhead line solutions, etc) developed and their journey from idea to reality. With several of the solutions being turned into marketed products.

We would constantly look out for untried, new, or emerging technologies that we could use to create innovative solutions. In the late 1980's fibre optics, microprocessors, mobile phones, navigation systems, wind & solar energy, desktop computers & software, etc all of which transformed what was possible.

Any new solution would need to be reliable and railway-proof. We always developed & tested prototypes before implementation. This would mean multiple iterations, spin-off projects (ie the dome), or projects that have to be closed down.

workplace

BR Research
BR Research – Railway Technical Centre, Derby
Centre of Excellence for Research into Science & Technology for the Future Railway Network

Specialist Facilities
- Laboratories (ie HV-300kv)
- Test Halls
- Engineering Workshops
- Lab/Test Vehicles (Promethius)
- Test Tracks (Old Dalby OHL)

BRITISH RAILWAYS RESEARCH DIVISION

Recognised internationally as one of the worlds leading centres of railway expertise. & employing 700+ scientist & engineers.

LABORATORIES

The centre had specialist facilities for r&d including laboratories, workshops, test halls & main line test vehicles.

Electrification Team - BRR

Electrification Section
Leading Edge Projects in support of Infrastructure & Vehicle aspects of Electrification Technology

A **Multi-skilled Team** of Scientific Engineers, with specialist knowledge in the following fields:
- HV Electrical (Equipment, Distribution etc)
- Current Collection (ie Dynamics)
- Mechanical (Design & Build)
- Instrumentation (Measurement)
- Electronics (Circuit Design)
- Computing (Programming)
- Production & Manufacturing (Workshop)

Producing Solutions, Technical Reports & International Papers

ELECTRIFICATION TEAM

A multi-disciplinary team of scientific engineers, with specialist knowledge in a range of fields. Working in small & flexible project teams.

QUALIFICATION

Degree in Mechanical Engineering (Hons) Part-time study

technical & scientific knowledge

RAILWAY ELECTRIFICATION

The electrified network in the UK uses a 25Kv AC Overhead Line and 750v DC Third Rail systems. Electric trains are preferred for major railway lines because electric trains are lighter, cleaner, cheaper, quieter, and faster to accelerate. They allow more trains to be run more efficiently and quickly.

OVERHEAD LINE SYSTEM

The purpose of the Overhead Line Equipment (OHE) is to supply electricity to moving trains. The system works by :

- Feeder stations transmit power to the OHE
- Overhead structures support overhead wires
- Power is collected by a sprung pantograph
- Flows through a transformer & controls
- To drive traction motors that turn the wheels

The principal aims of the OHE designer are to keep the contact wire as stationery as possible so that power can flow uninterrupted to the train, and minimising wear of the system. To achieve this the contact wire is tensioned and runs in a zig-zag path above the track, but rises and falls for level crossings and tunnels.

The challenge for Electric Traction is collecting current reliably at high speed, this requires an understanding of the dynamic behaviour of both the pantograph and overhead line system. The system relies on a carbon strip (that has to be wearable) to stay in contact with a lightweight copper wire. Certain conditions can combine to create dewirements and have a major impact on the train service.

electrification systems

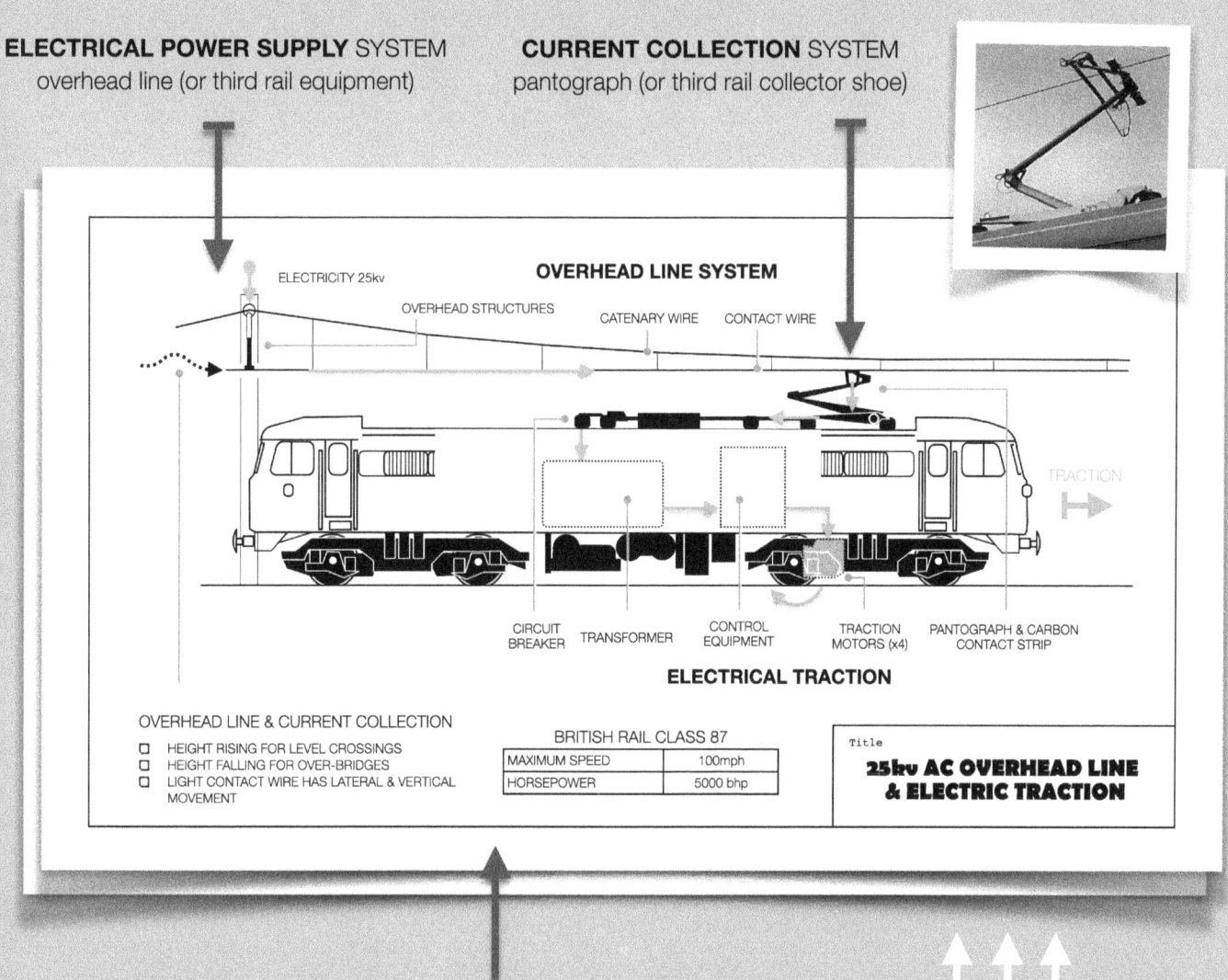

research : electrification

The pantograph and overhead line projects described here produced significant reliability gains and this is the story of their 3 year development. Each of the radical projects followed a rigorous approach to understanding causes and innovation, with reports being published to record discoveries and share lessons internally.

OVERHEAD LINE CURRENT COLLECTION

PANTOGRAPH DAMAGE ANALYSIS

improve train service performance

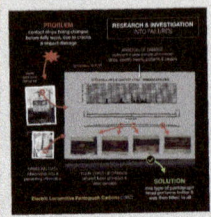

pantograph faults can lead to serious damage to the electrification system, train delays & loss of revenue

analysis of contact strips from hundred of service locomotives identified some causes

OVERHEAD LINE EQUIPMENT (OLE)

Is the assembly of masts, gantries & wires that supply power to make electric trains move. The power is transmitted from a contact wire to the train by a sprung pantograph.

PANTOGRAPHS

Pantographs are vulnerable to damage because of their necessarily low mass construction and the large impacts received from debris or badly adjusted or displaced fitting in the overhead line equipment.

PANTOGRAPH CONTACT STRIPS

Pantographs use carbon collector strips to minimise the wear on the overhead wire. However, carbon can be broken by large impacts and can then cause damage to long lengths of overhead line equipment before the fault can be detected.

WORKING WITH THE RAIL BUSINESS OPERATORS

Contact at all levels with railway engineers and operators gave us first-hand insight into both practices and problems. Working with the operating railway meant that real data could be collected, solutions tested, successfully implemented, and demonstrate real impact on business performance.

PANTOGRAPH INSPECTION AT STATIONS

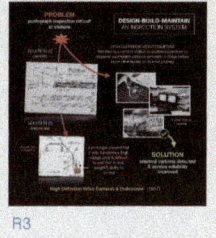

difficult to inspect pantographs in service

PANSEE product (camera system)

closed circuit television system provides an additional tool to check pantograph condition

SYSTEM ANALYSIS & PROJECTS

the whole system, its components, failure modes & links to development projects/solutions

pantograph faults

pantograph in-service performance not monitored

TRACKSIDE PANTOGRAPH CONDITION MONITORING SYSTEM (PANCHEX)

Test sites created on the West & East Coast mainline. Equipment then developed to daily measure the performance of all pantographs in service.

RESEARCH REPORT

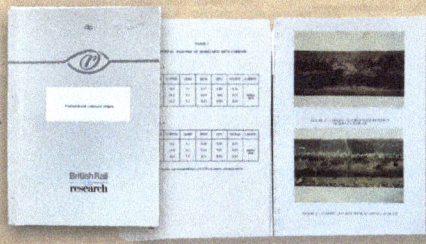

study of pantograph carbon contact strips, failure modes, material properties, exploring alternative materials & fault detection/ protection methods

overhead line faults

pantographs are being damaged by overhead line equipment

141

development : pantograph

The VISION : to develop a system to continuously monitor the condition/health of in-service pantographs and report or action defective pantographs.

MEASURING CONTACT WIRE UPLIFT

instrumentation installed on the live overhead line & data transmitted to

IDENTFYING PANTOGRAPH TYPE & ORIENTATION

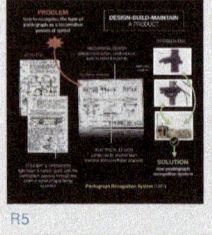

pantograph type & direction detected by a beam of infra-red light being cut by the passing pantograph frame

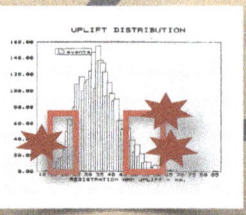

defective & deteriorating pantographs detected early & rectified by maintenance depots

leading to significant improvement in the condition of pantographs in service

PANTOGRAPH LAB TESTING

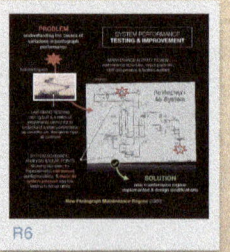

pantograph bench tests identified system weaknesses, leading to component improvements & a new maintenance regime

PANCHEX product (condition monitoring system) commercialised & installed across Britain & in Spain

PANCHEX : PRODUCT

The condition of pantographs & their behaviour is assessed by measuring the uplift & lateral acceleration of the contact wire caused by a passing pantograph. The information is then used to detect & rectify defective pantographs.

& overhead line

USING NEW TECHNOLOGIES

The projects made use of emerging technologies including, fibre-optic instruments, micro-processes, mobile communications & navigation systems.

The VISION : to develop a system to monitor/report the condition of the overhead line across the network, that could be fitted to any passenger train and detect/report overhead line faults.

OVERHEAD LINE CONDITION MONITORING SYSTEM (OLIVE)

impact detection

fibre optic accelerometer mounted on the pantograph

overhead line system only inspected once or twice a year

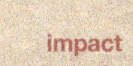

fibre-optic cable insulator

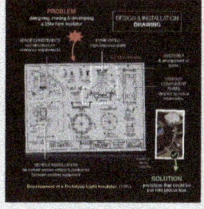

optical fibre is brought down through an oil filled porcelain insulator

instrumented test coach checks overhead line parameters

initial concept for a locomotive fitted system, using new & emerging technologies

equipment installation

data processing & performance monitoring

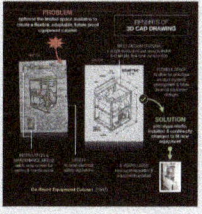

OLIVE : PRODUCT

Service locomotives monitoring the condition of the overhead line with an instrumented pantograph. A fibre-optic accelerometer measures the severity of impacts between the overhead line & pantograph,. The data is the need to detect & locate overhead line defects.

electrical, electronic, & computer system developed by the other project team member

fibre optic data processing, together with navigation & mobile communication equipment housed in a cabinet on the locomotive

OBSERVATION DOME

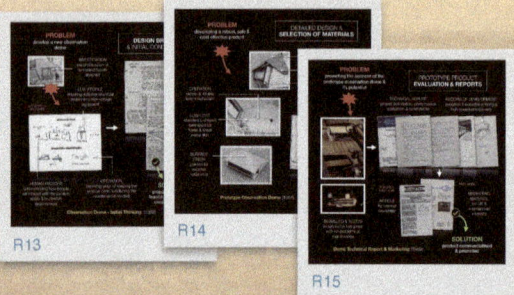

drivers can't inspect potentially damaged carbons

article written about the spin-off benefits for train driver fault detection

how to calibration the system & matching actual behaviour to data collected

dome design, made, installed & tested

PANDOME product & research report

CONDITION MONITORING OF NETWORK

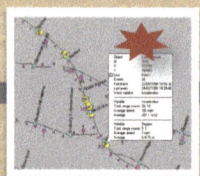

defective overhead line identified

the processed outputs gave a continuous 'health check' of the state of the overhead lie system with problems being identified & rectified

OLIVE product

equipment installed on several service locomotives & ran successfully for several years

mechanical vibration testing & additional improvements to make the system rail proof & reliable

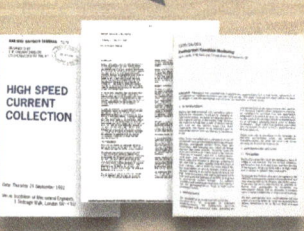

location identification so satellite navigation system fitted

real-time communication so mobile phone fitted

promotional video created for customers

learning shared with papers written & presented at several International conferences & seminars

other projects

Here are three other projects exploring technical innovations that could help shape possible future electrification systems.

OVERHEAD LINE MAINTENANCE

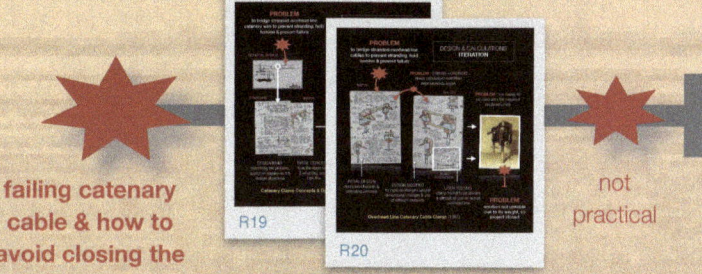

failing catenary cable & how to avoid closing the line to repair it

not practical

developed a prototype catenary clamp & tested

HIGH SPEED THIRD RAIL CURRENT COLLECTION

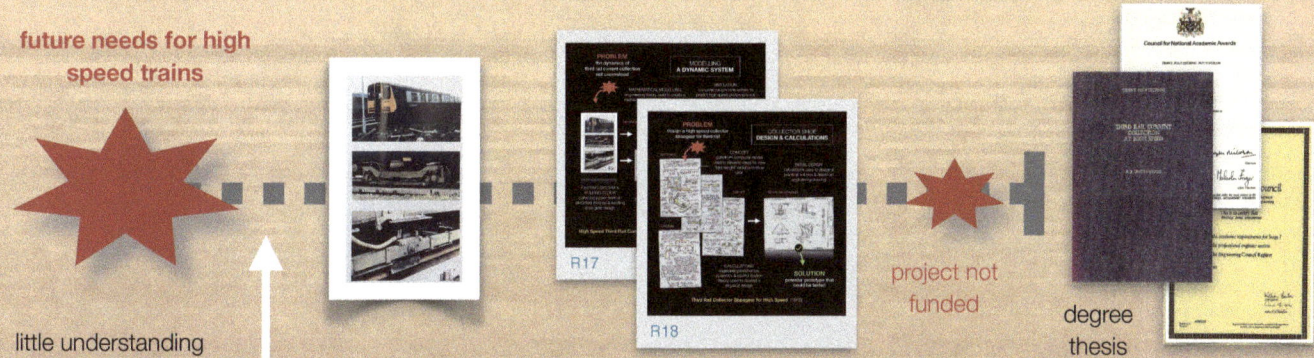

future needs for high speed trains

project not funded

degree thesis

little understanding of third rail current collection

creating a theoretical model & computer programme to simulate the third rail current collection system & develop possible shoe gear design

completed an honours degree in Mechanical Engineering & became an associate member of the IMechE

THIRD RAIL CURRENT COLLECTION

As railways strive for higher speeds, an understanding of modelling of high speed current collection performance would be needed. The dynamics of third rail & collector shoe interaction was modelled, & future shoe-gear design concepts explored.

BEYOND THE PANTOGRAPH

explore alternatives to the pantograph

new radical concepts explored

no business case

145

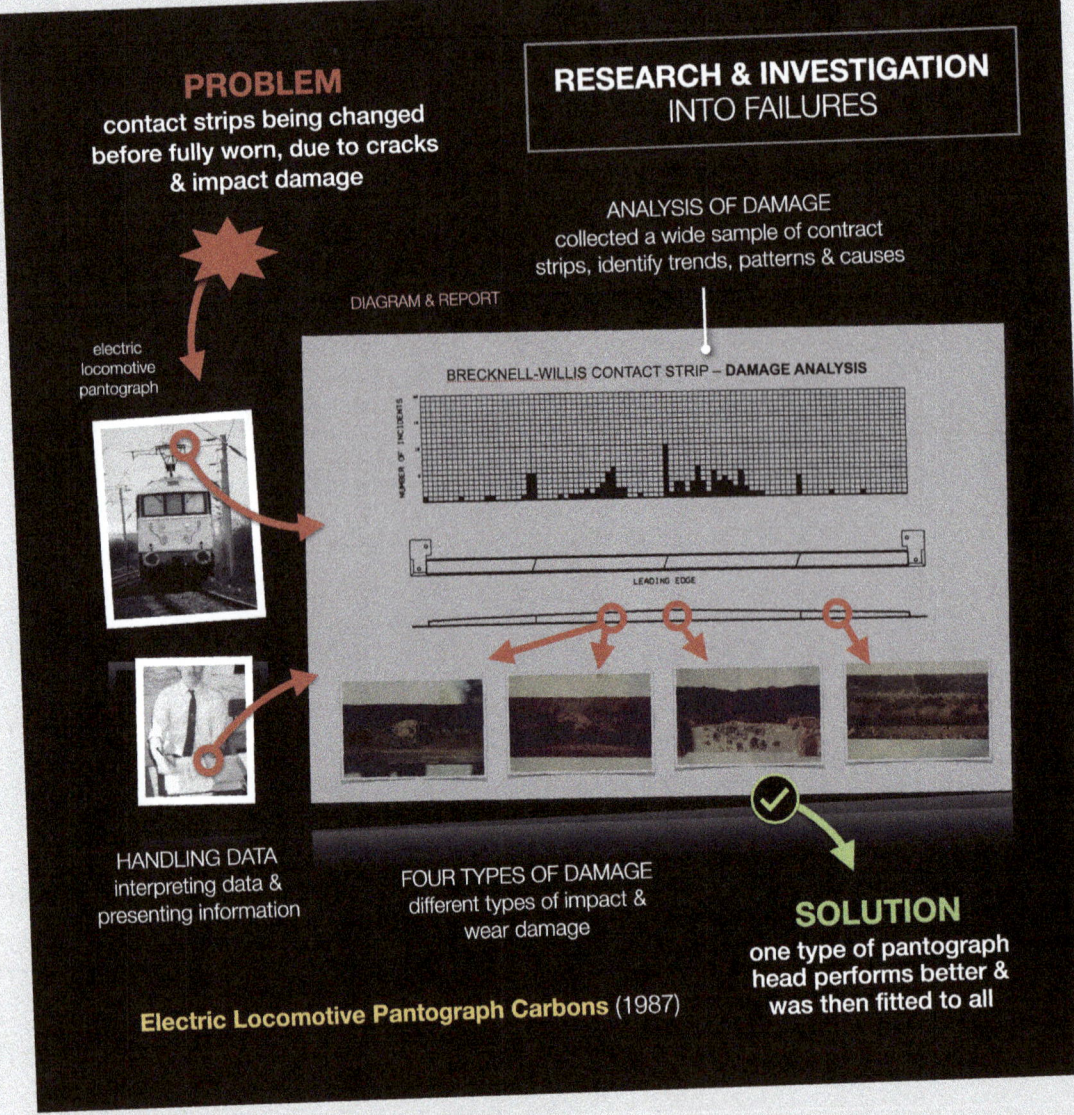

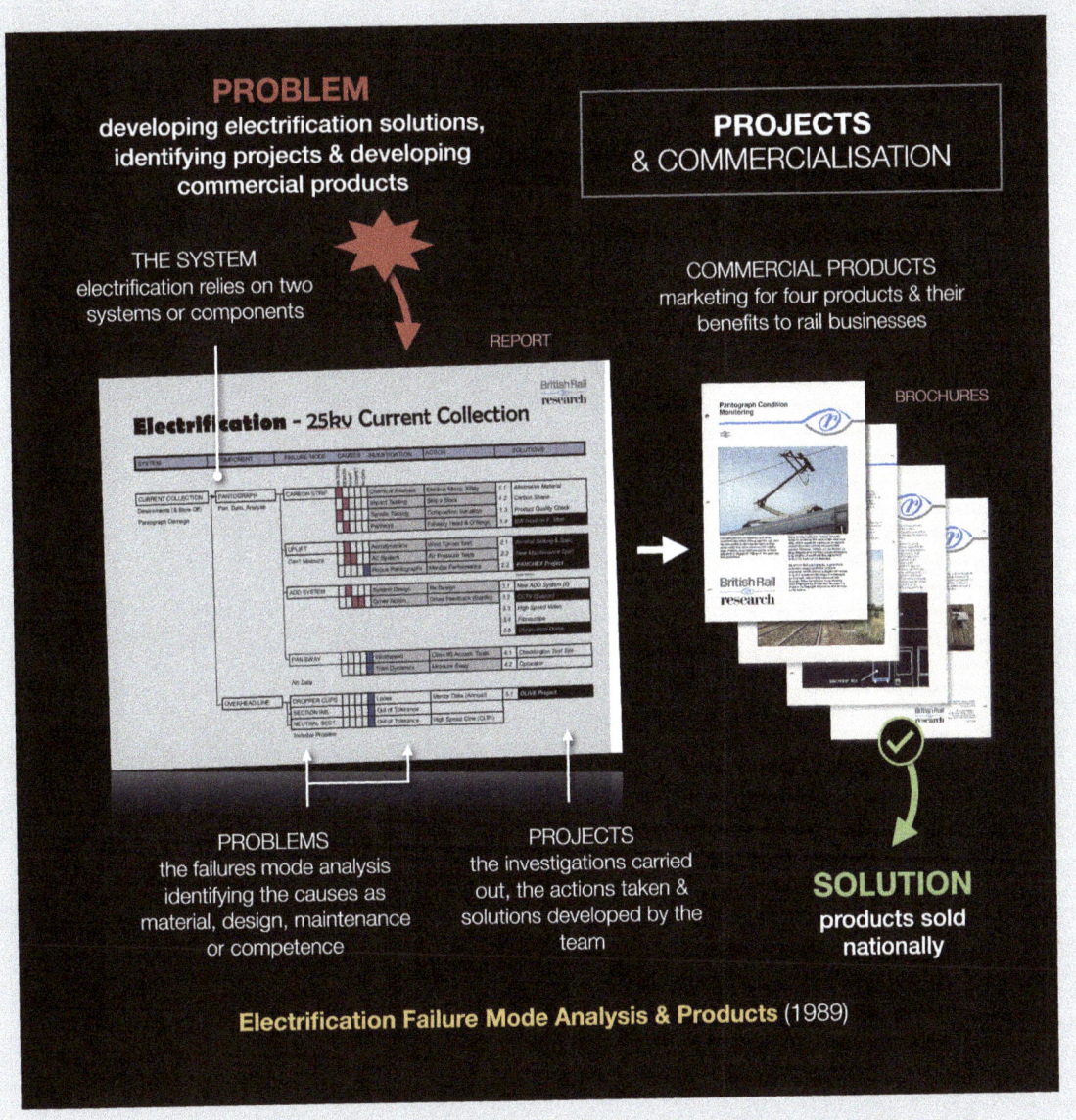

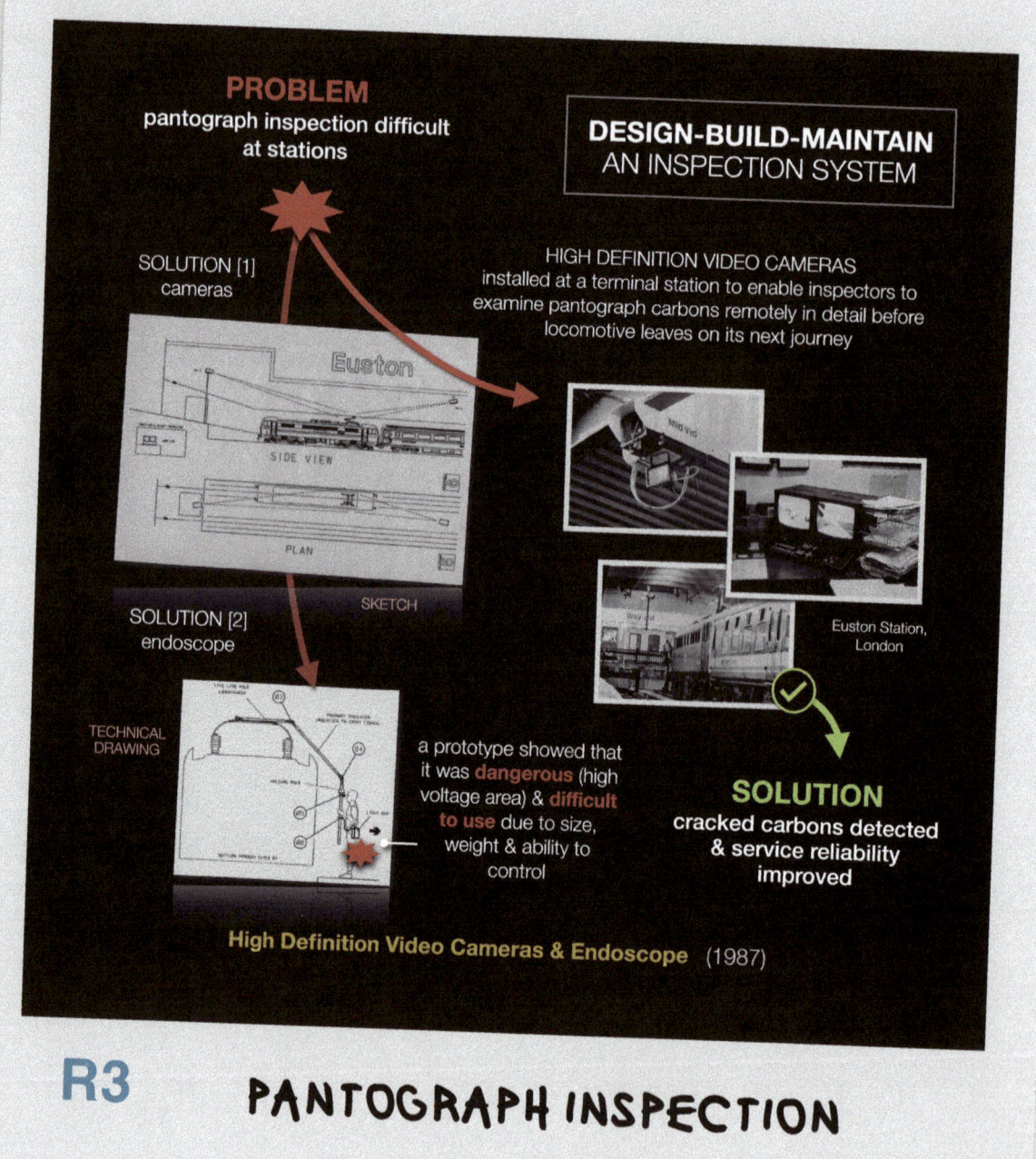

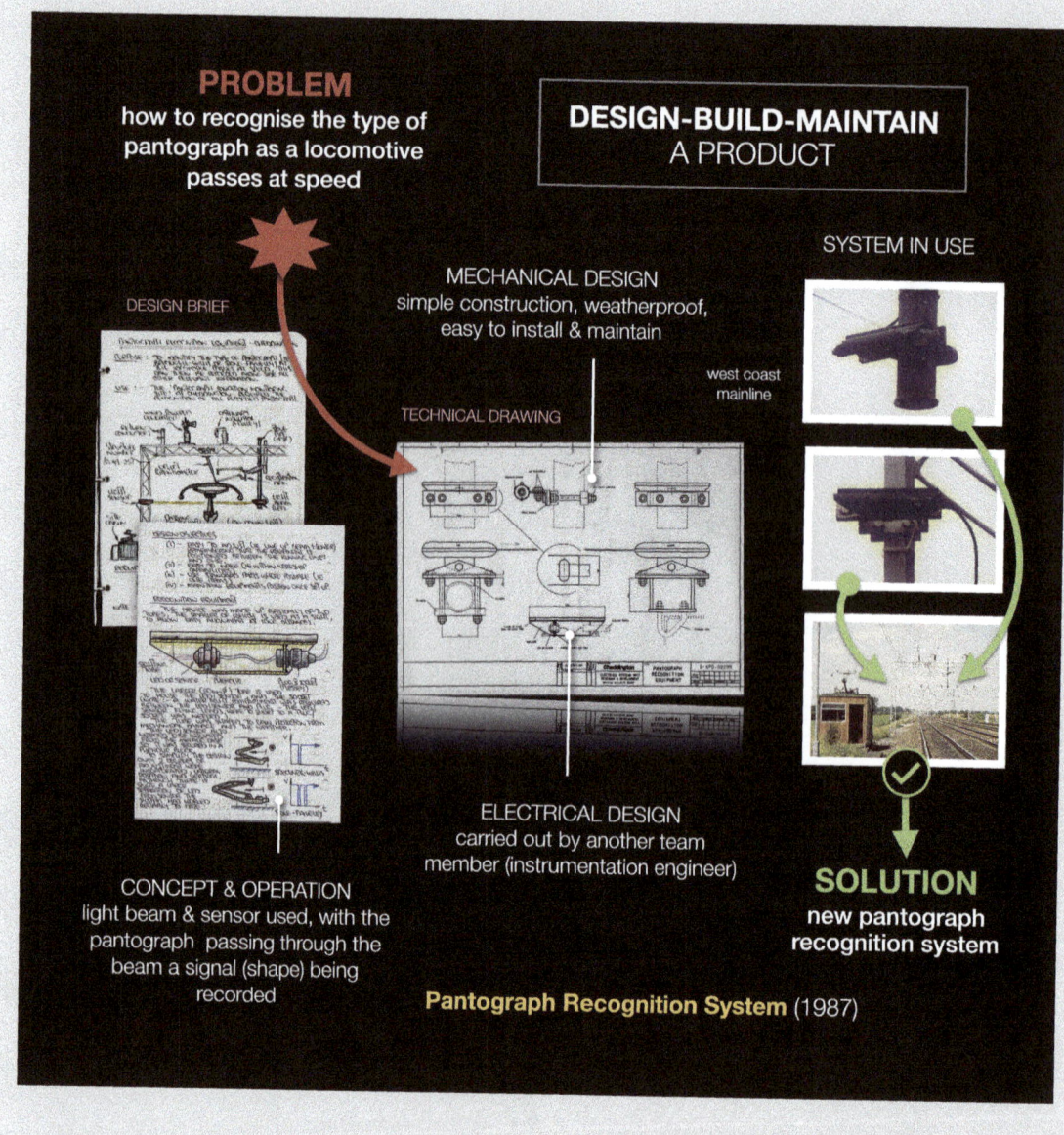

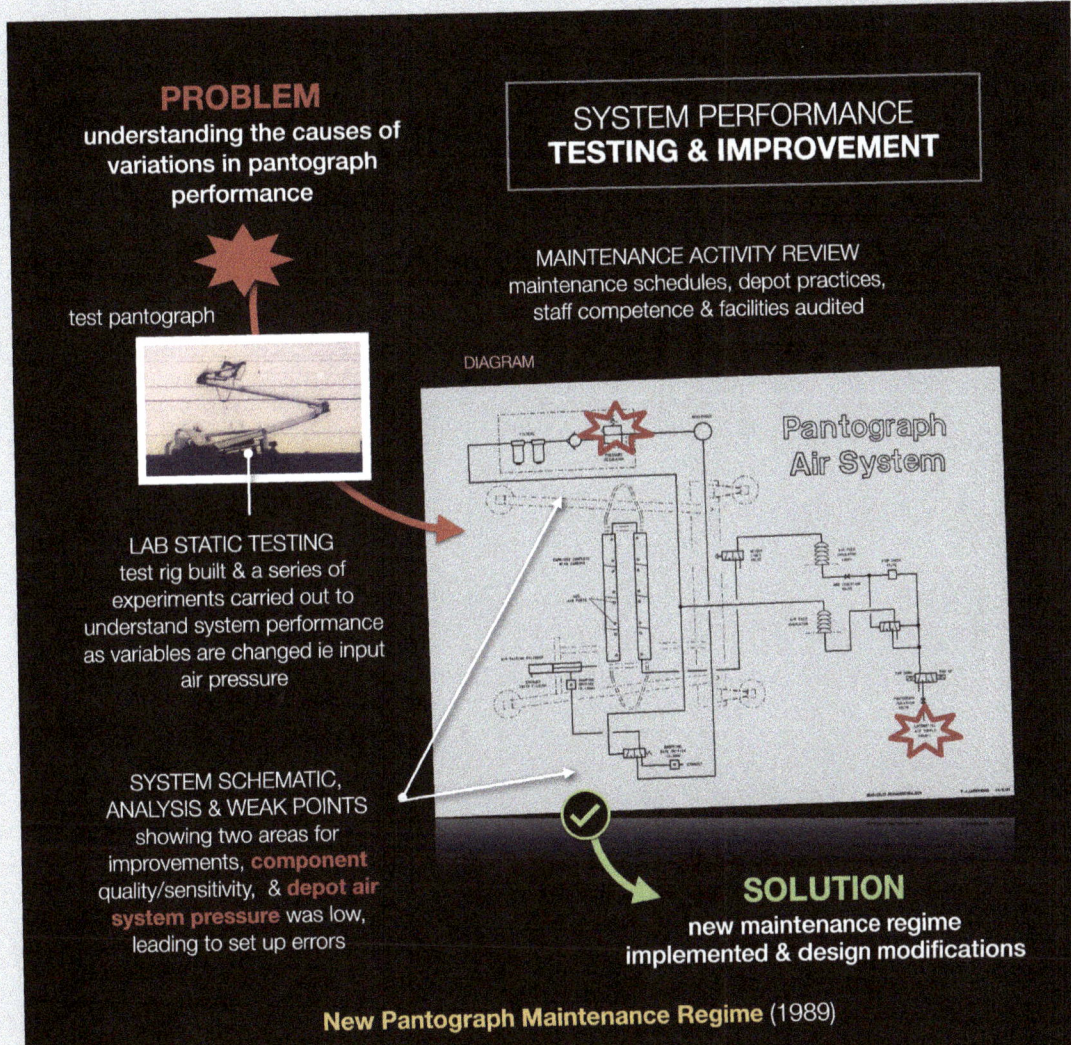

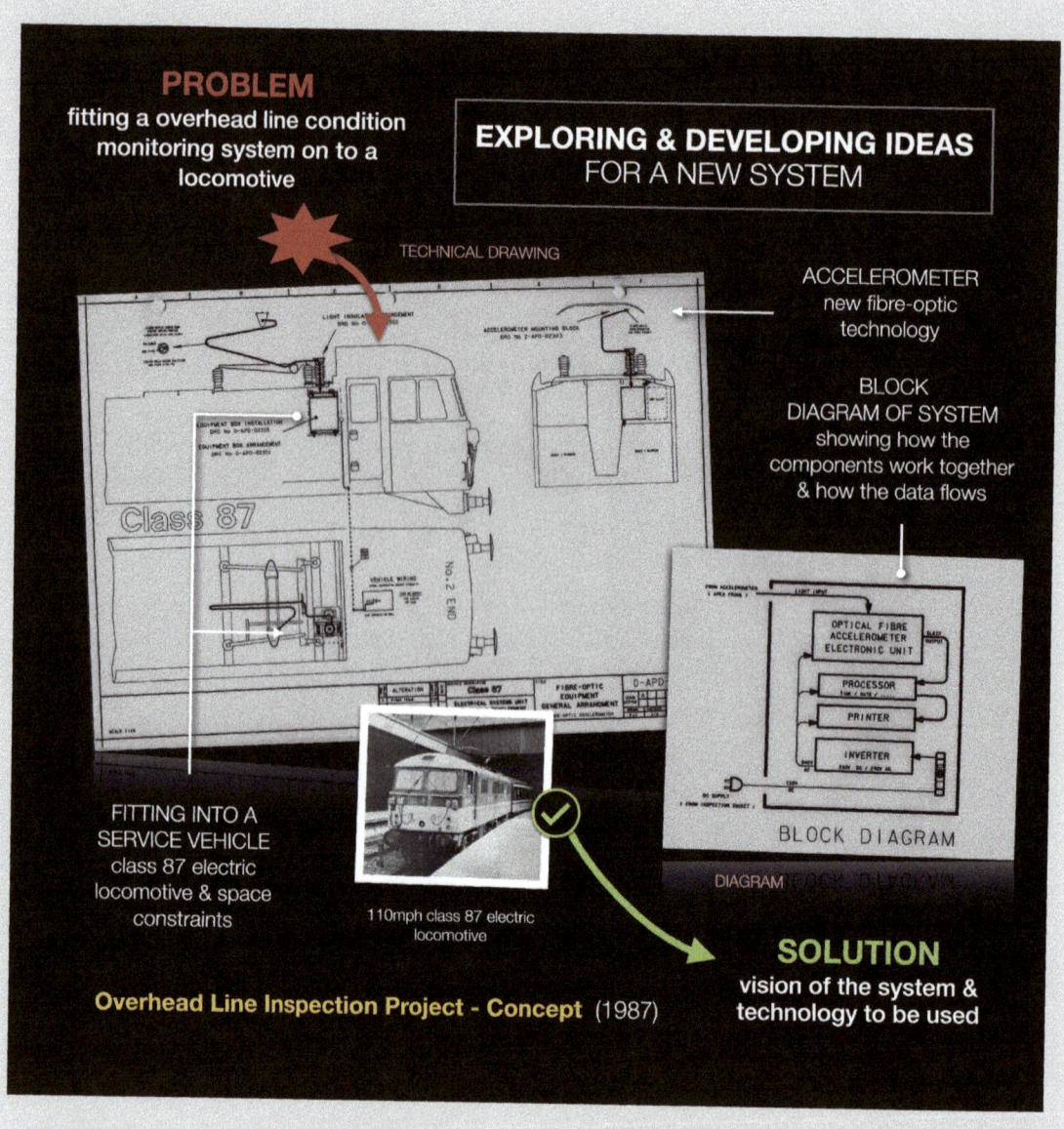

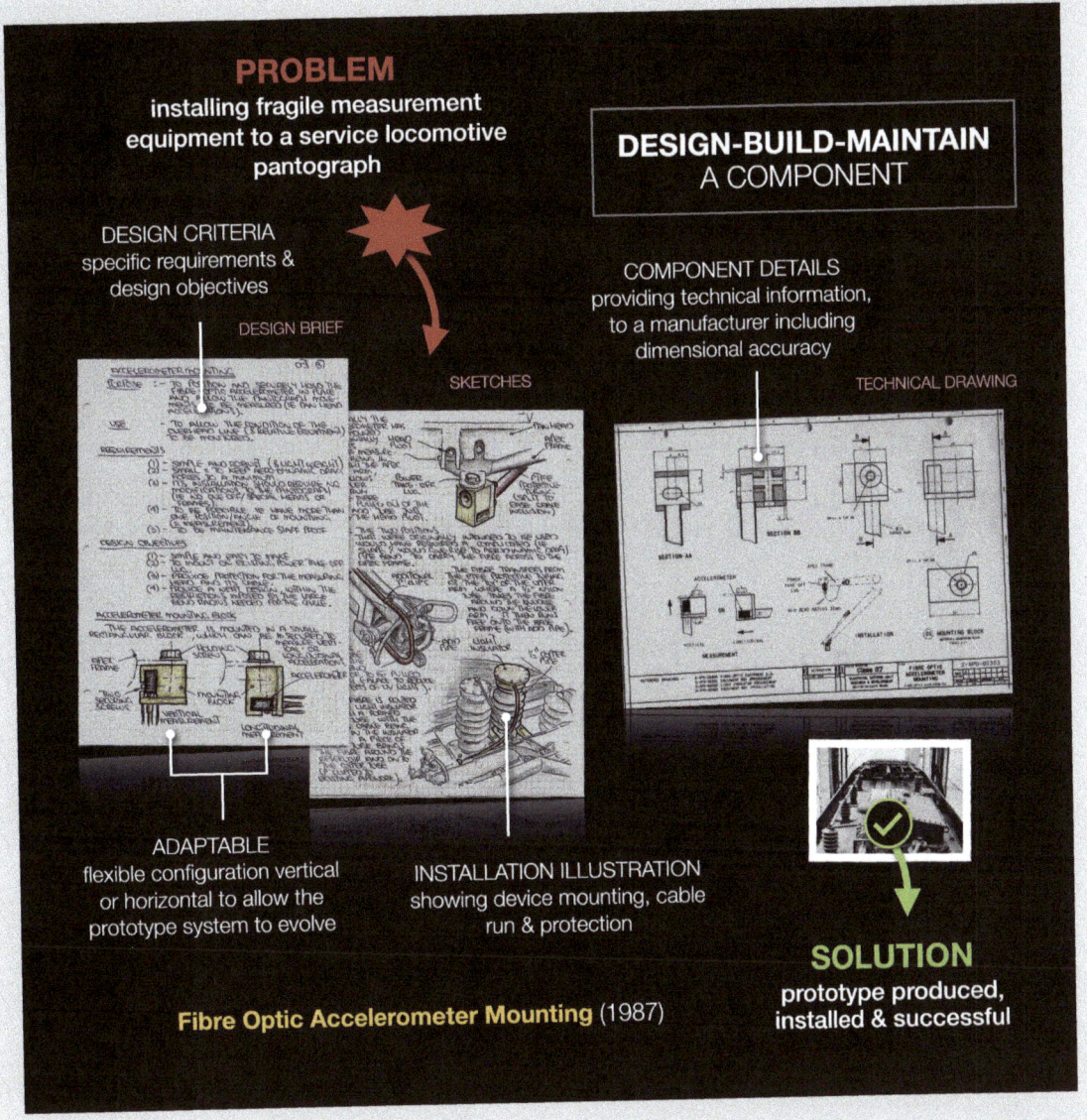

R8 OVERHEAD LINE MONITORING [2]

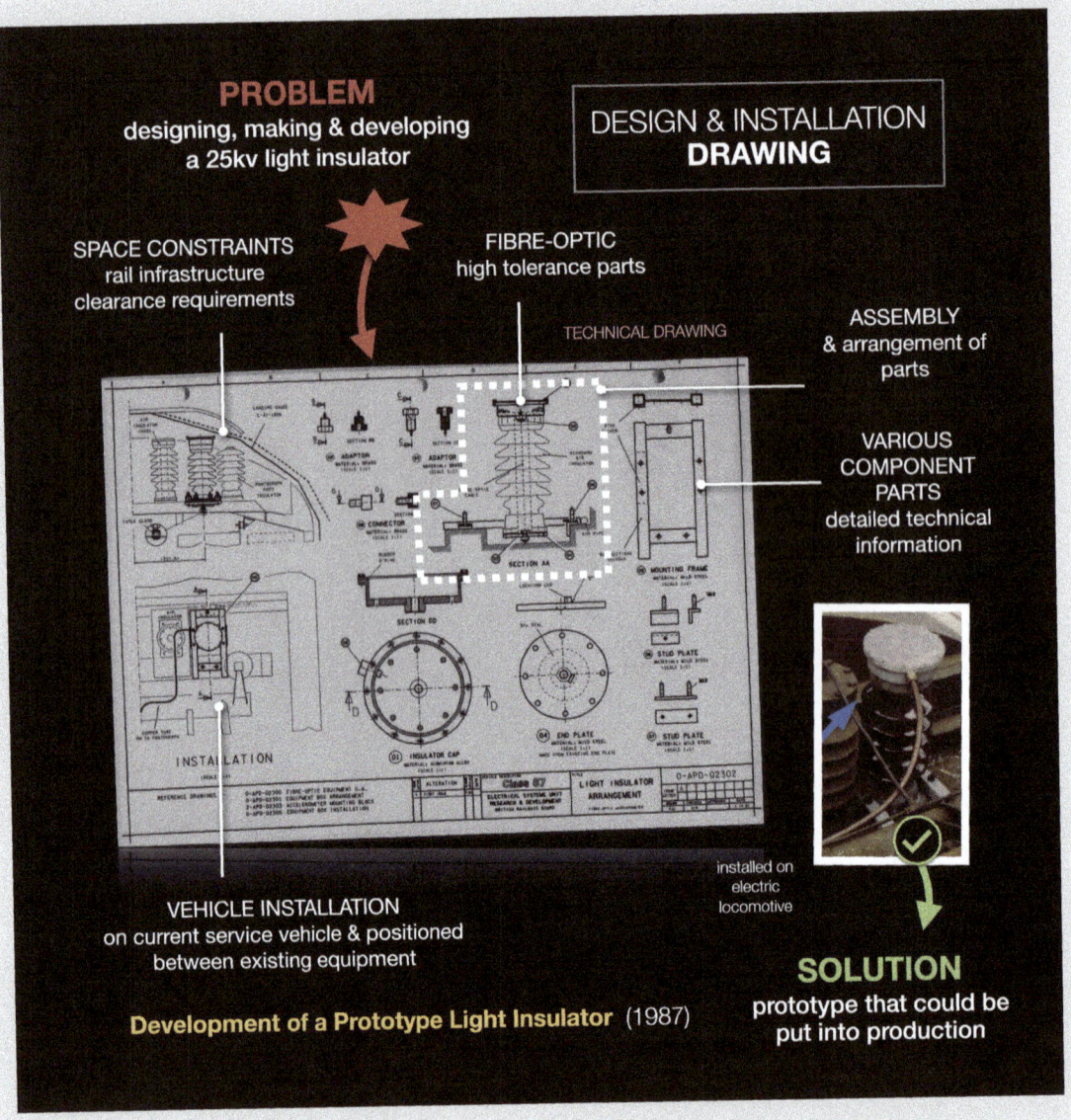

R9 OVERHEAD LINE MONITORING [3]

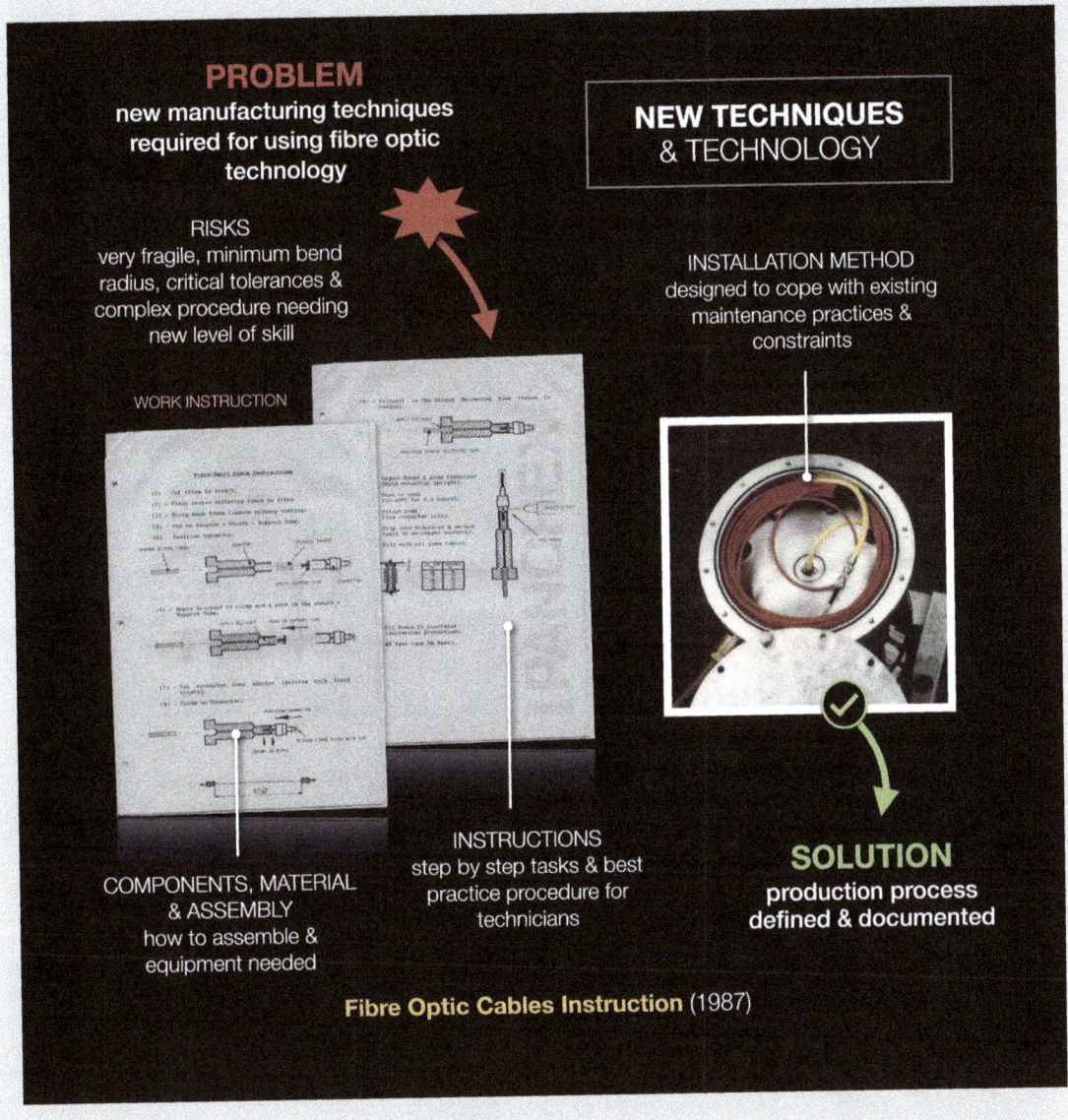

R10 OVERHEAD LINE MONITORING [4]

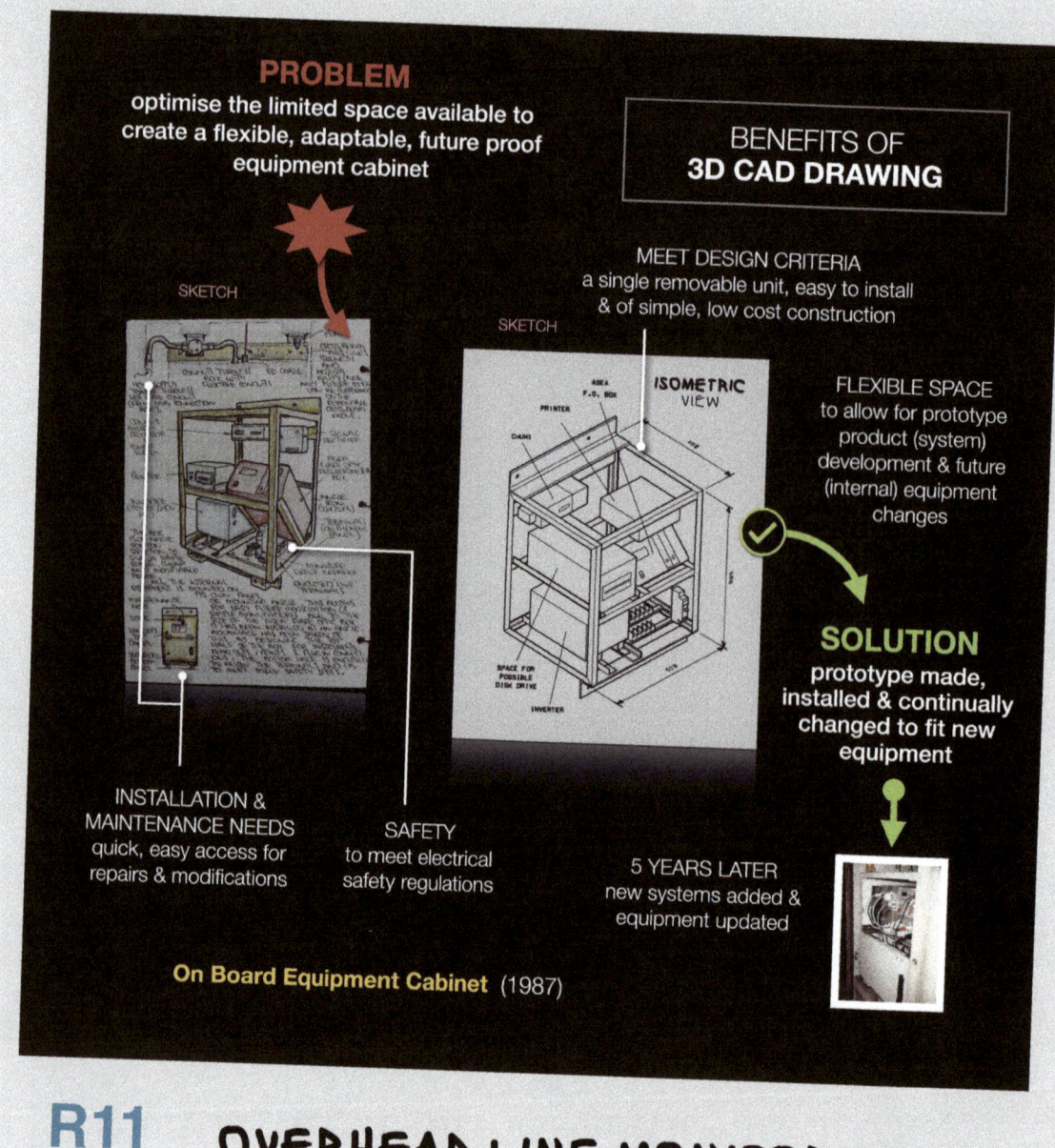

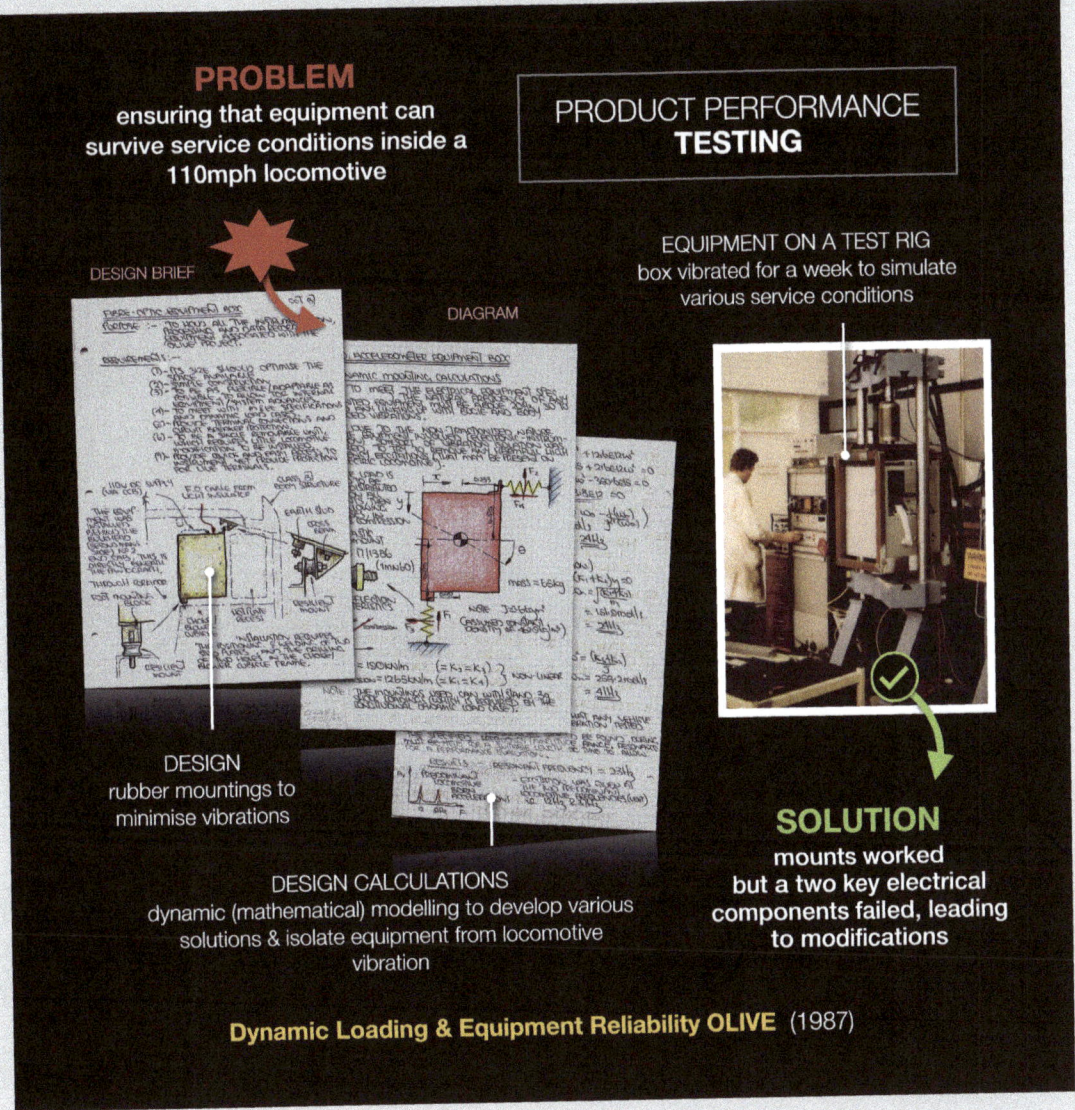

R12 OVERHEAD LINE MONITORING [6]

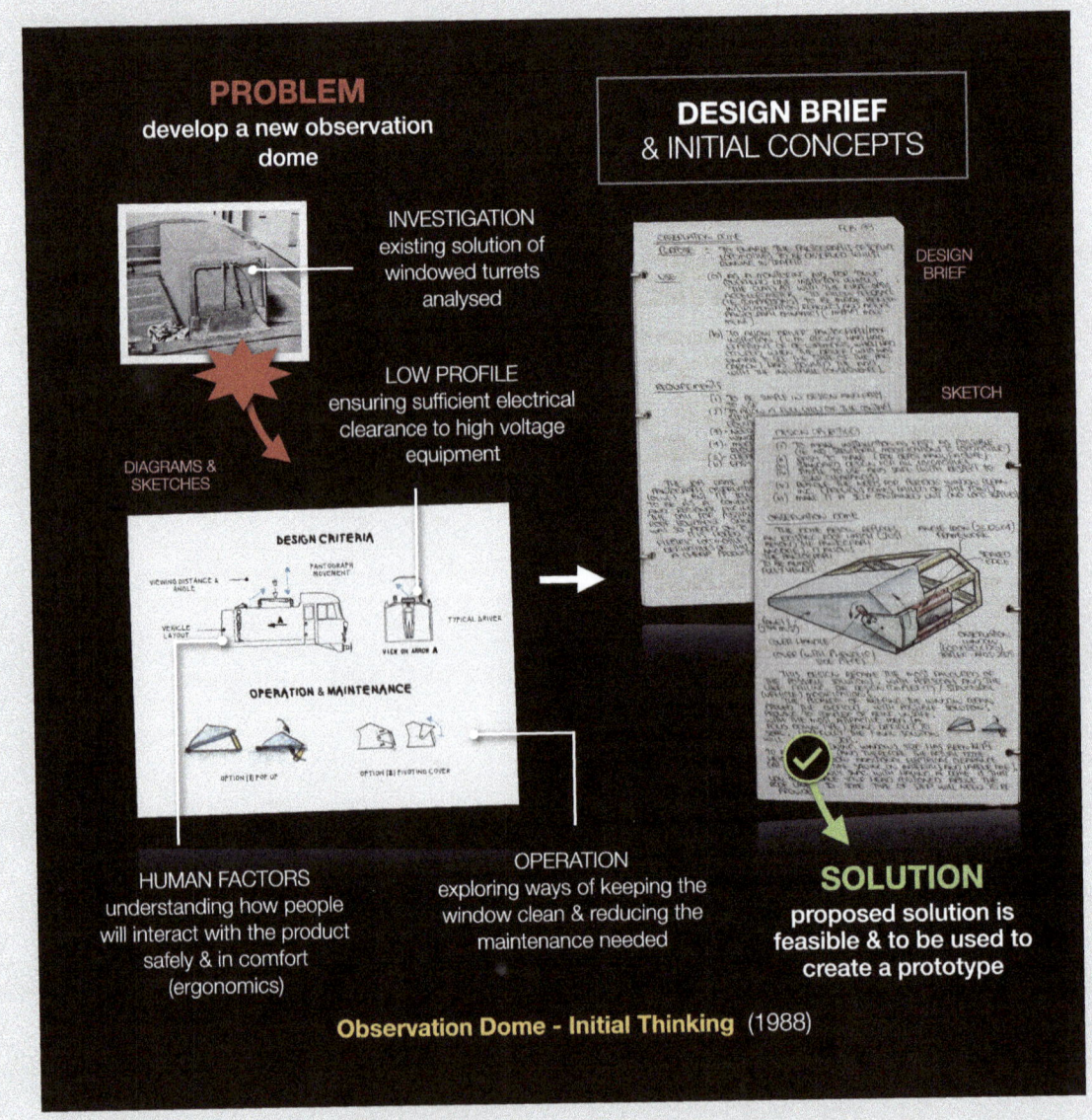

Observation Dome - Initial Thinking (1988)

R13 OBSERVATION DOME [I]

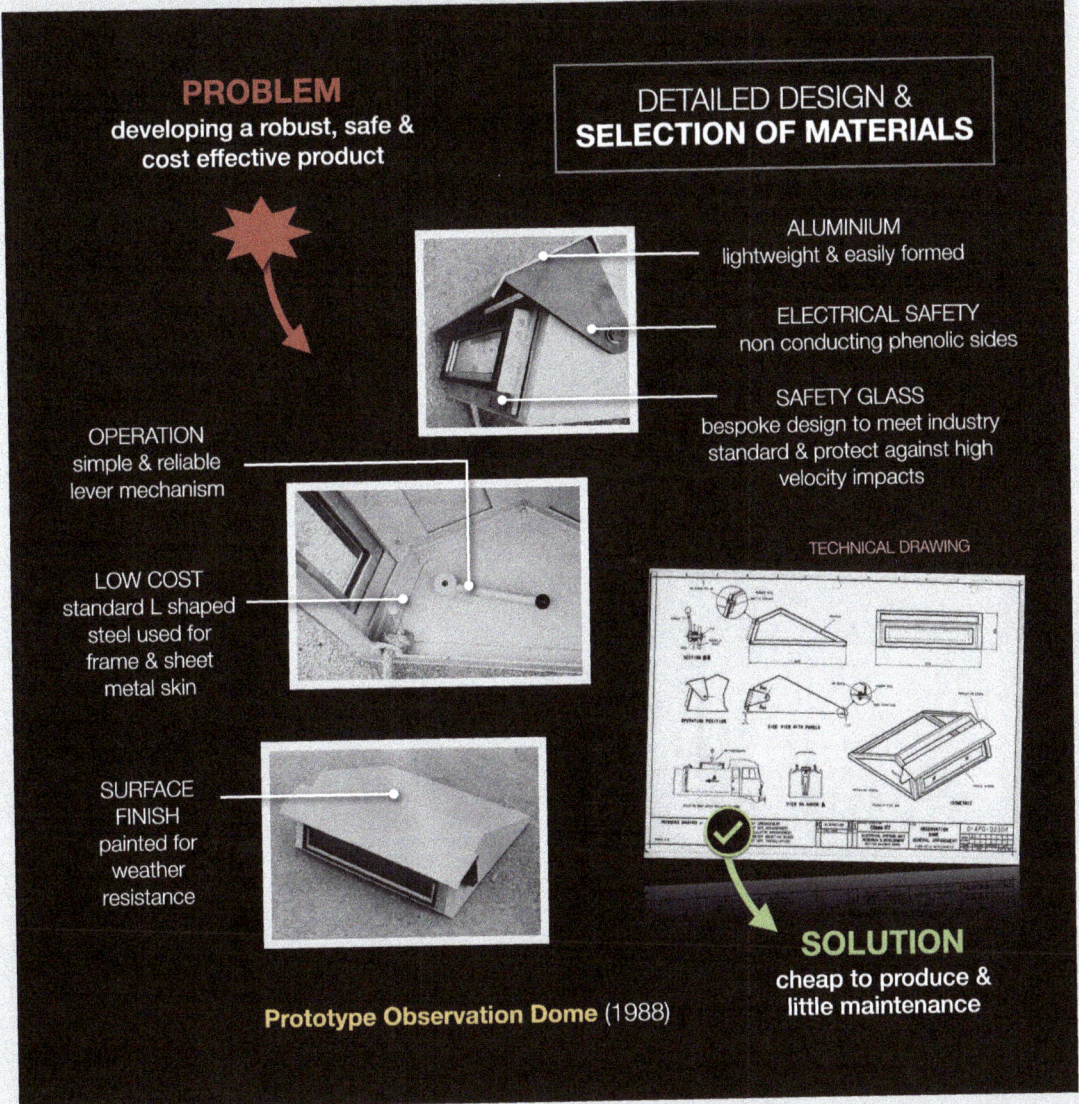

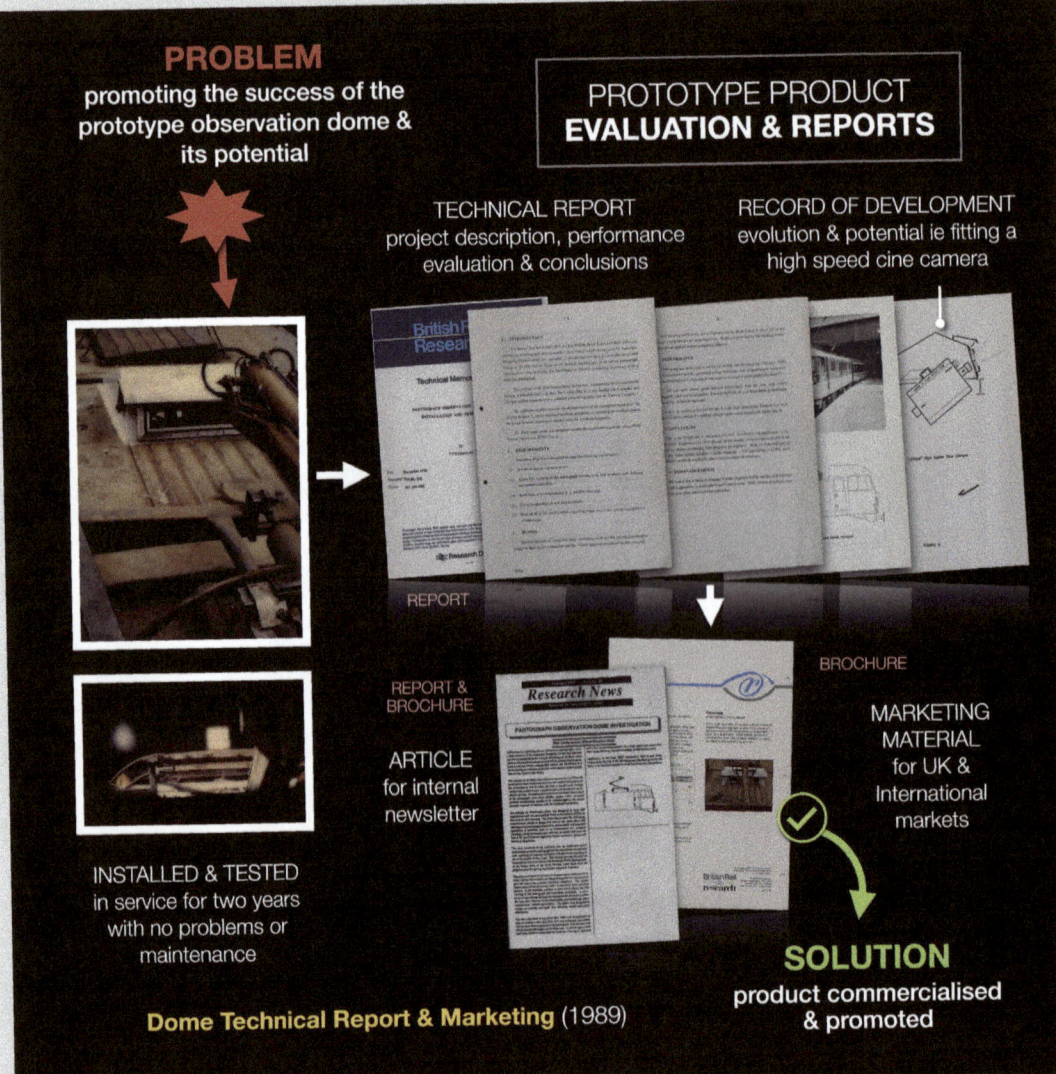

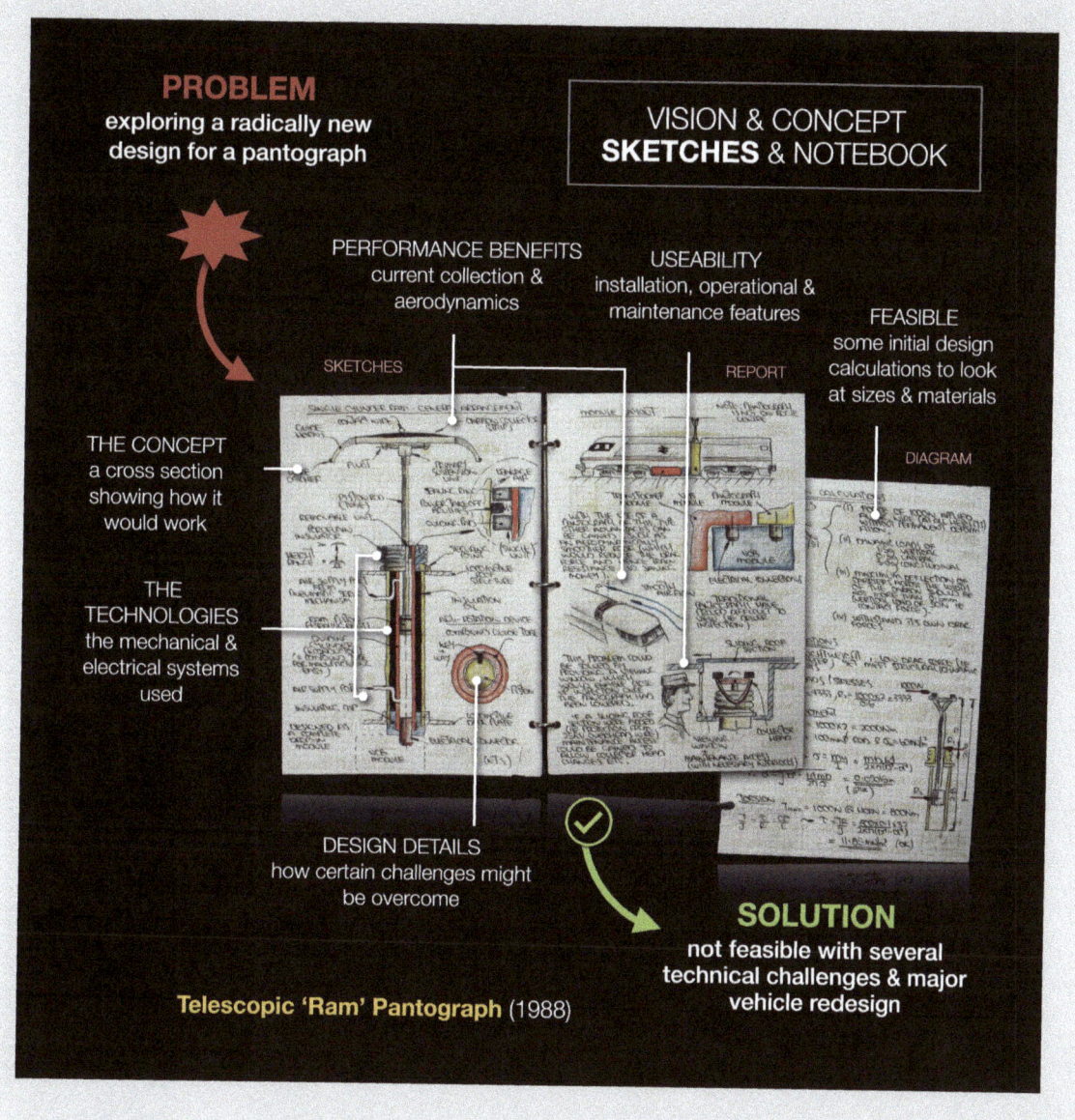

MODELLING A DYNAMIC SYSTEM

PROBLEM
the dynamics of third rail current collection not understood

MATHEMATICAL MODELLING
engineering theory used to create a mathematical model of the dynamic mechanical system

SIMULATION
computer programme written to predict high speed performance & validated against practical data

DIAGRAMS

REPORT

EXISTING SYSTEM & ROLLING STOCK
collecting power from an electrified third rail & existing shoe gear design

INPUT : RAIL PROFILE
the conductor rail behaviour also modelled & dynamic profile calculated

SOLUTION
work shows that there is a limit to performance of existing equipment & identified potential improvements

High Speed Third Rail Current Collection (1989)

R17 — THIRD RAIL CURRENT COLLECTION [I]

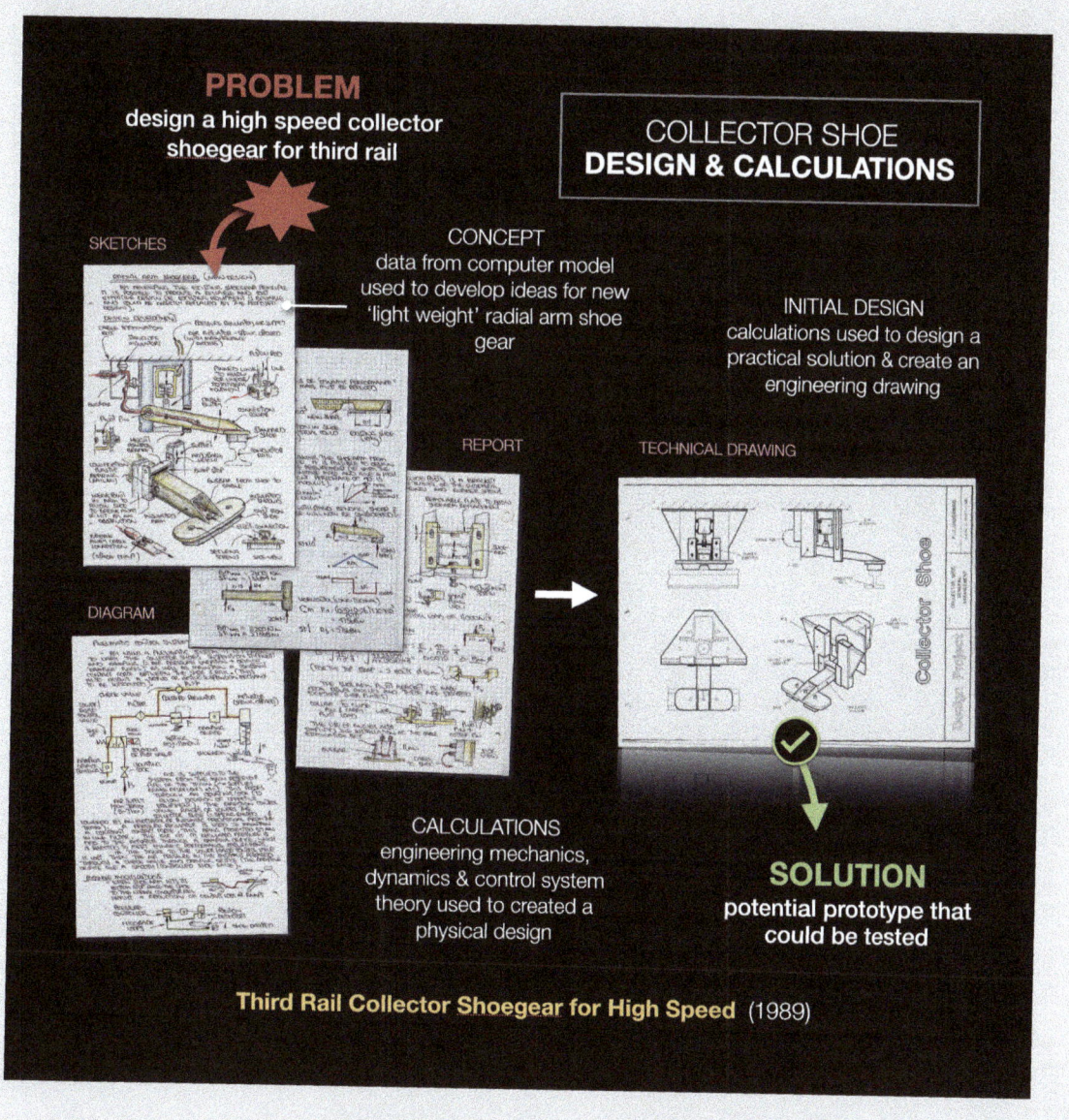

THIRD RAIL CURRENT COLLECTION [2]

R18

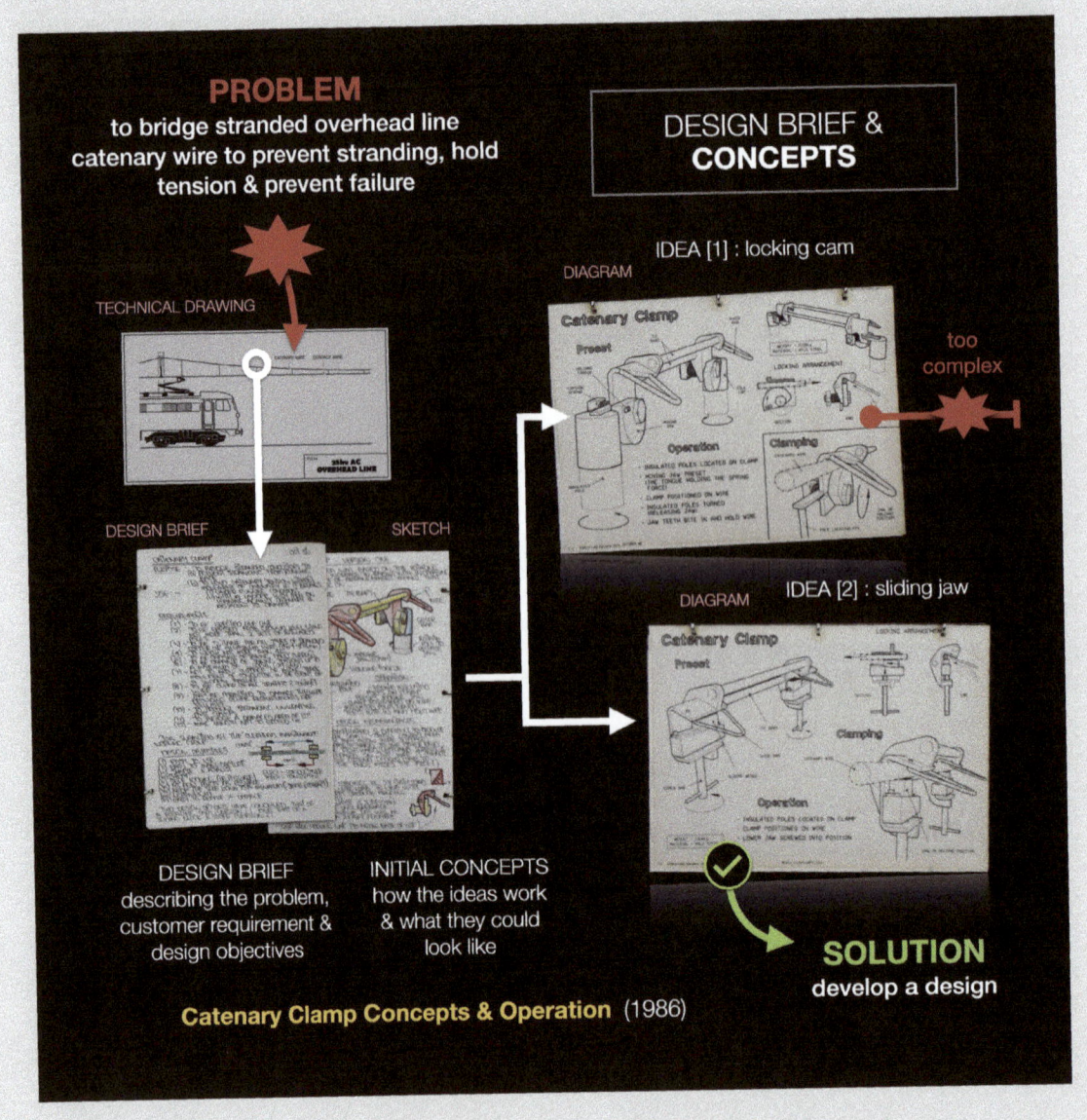

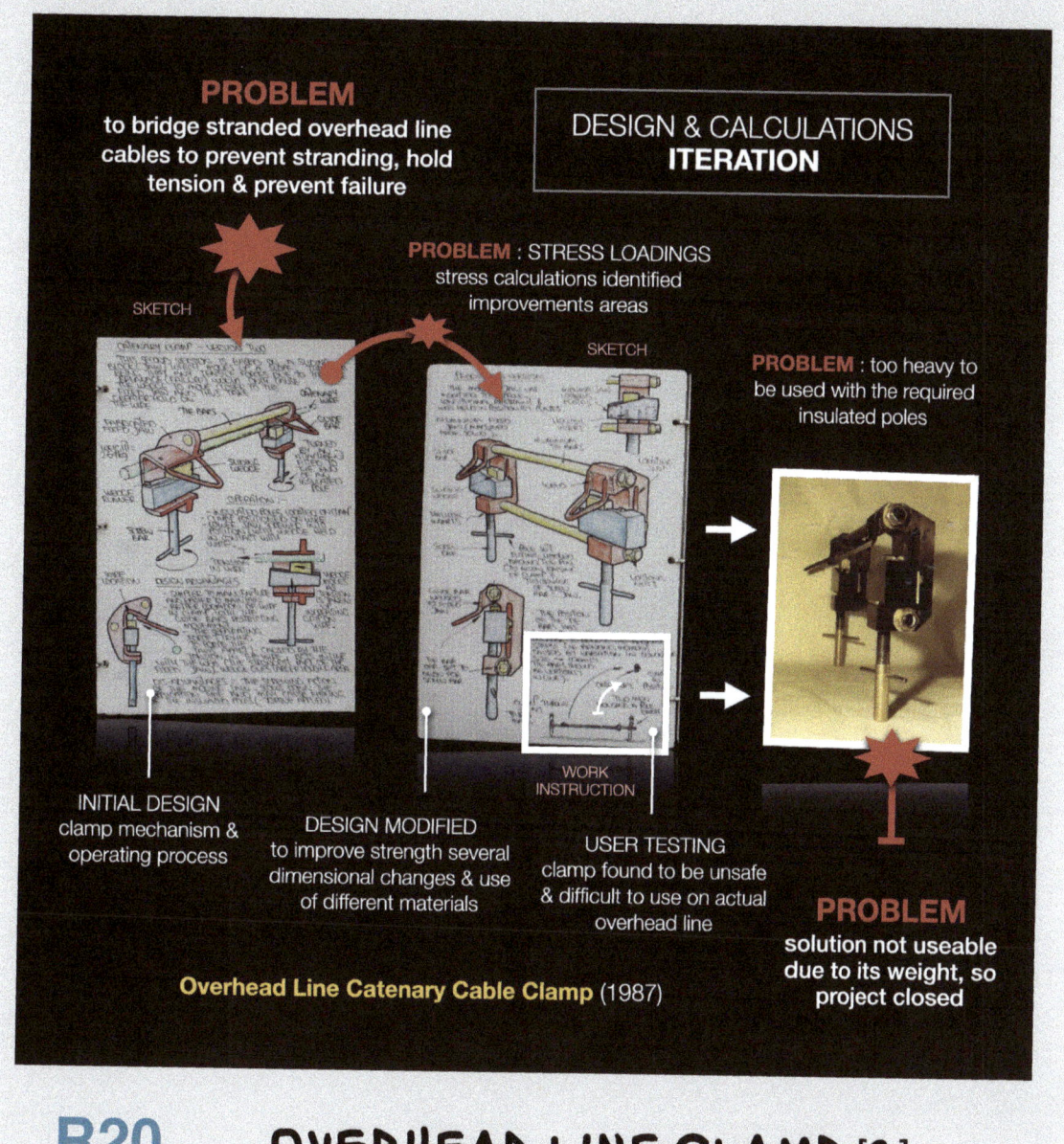

INSIGHTS on research,

SYSTEMATIC INVESTIGATION
a rigorous approach to cause analysis & theoretical understanding of the whole system to establish the facts

DISCOVERY
the facts help to define the problem & can lead to new scientific theories or insights

IDENTIFYING OPPORTUNITIES
exploring possible new ways, and solutions & generating ideas

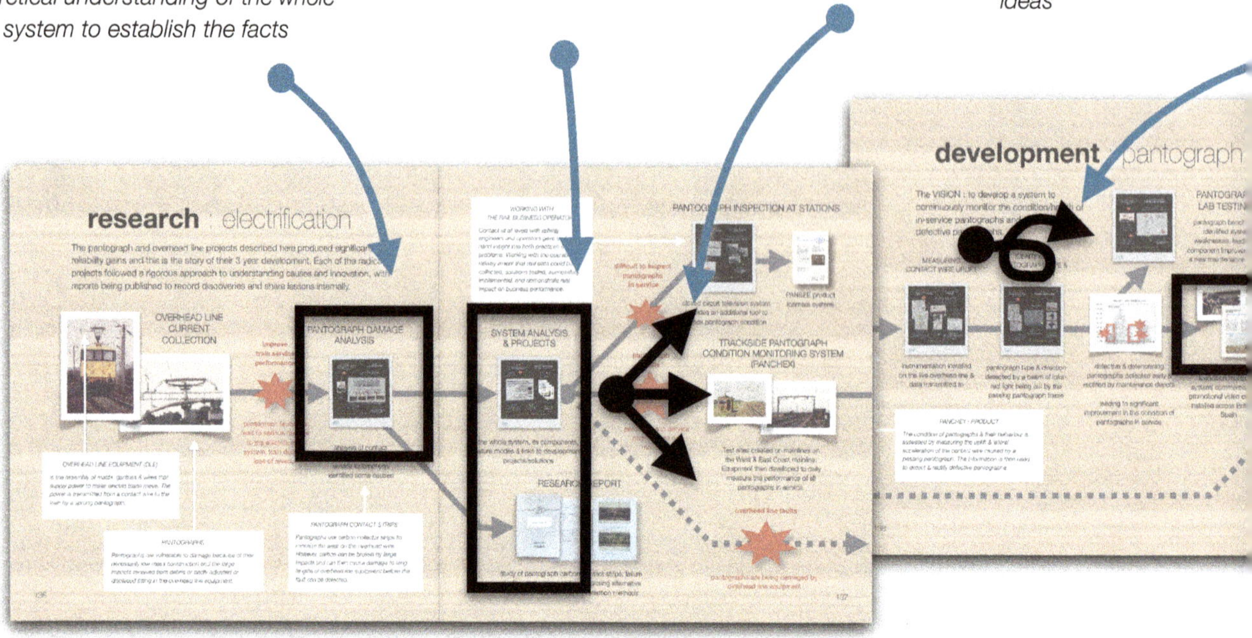

RESEARCH TIPS

- ☑ **systematically investigate** to establish the facts & problem
- ☑ this often leads to the **discovery** of something
- ☑ then start **identifying opportunities**

development & innovation

CONCEPT DEVELOPMENT
designing solutions, exploiting new technologies & developing prototypes

TESTING + VALIDATION
field trials, overcoming difficulties, reporting findings & projects continued or closed down

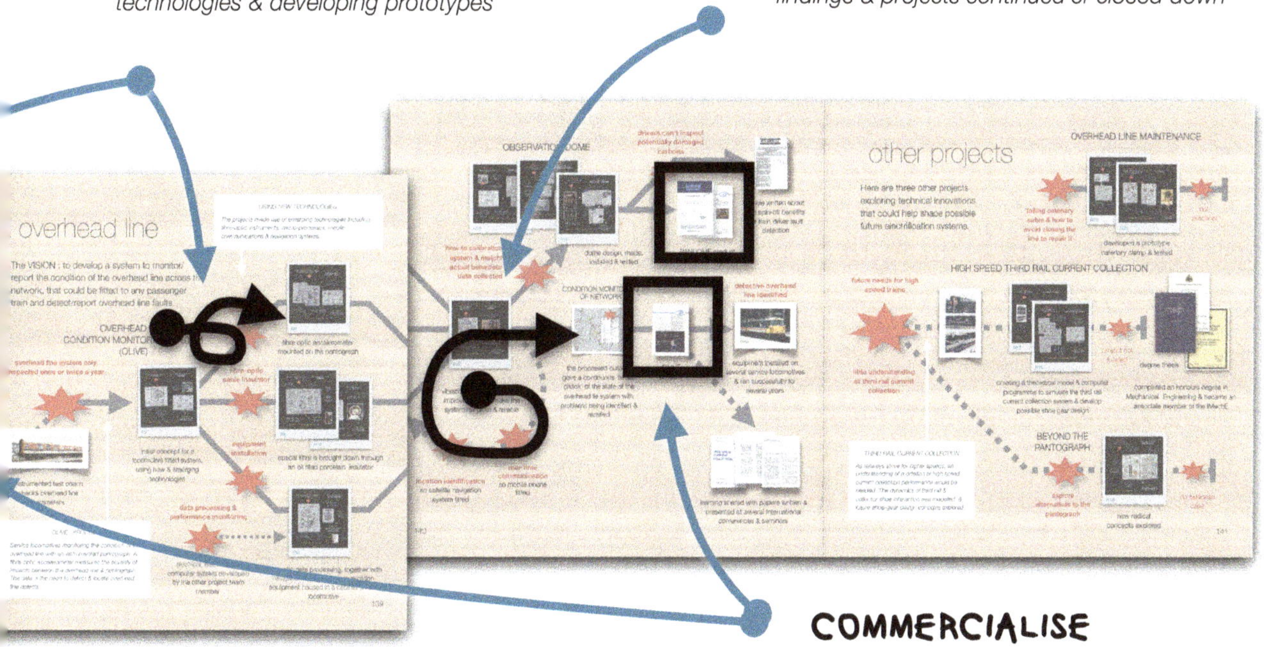

COMMERCIALISE
*product production, marketing & selling.
r&d is about putting money into ideas, to turn ideas into money (or savings)*

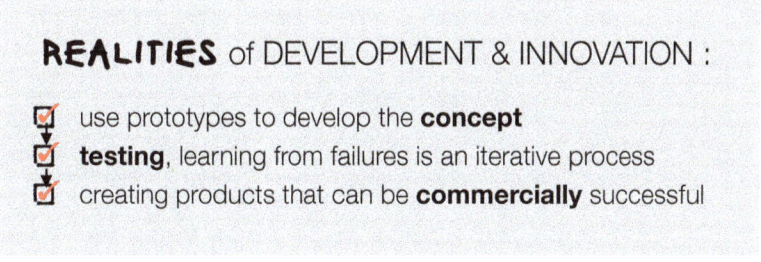

REALITIES of DEVELOPMENT & INNOVATION :

- ☑ use prototypes to develop the **concept**
- ☑ **testing**, learning from failures is an iterative process
- ☑ creating products that can be **commercially** successful

transferable skills

INVENTING THINGS

Working in research gave me an understanding of the difficulties involved in **putting innovation into practice**. The iterative process I learned would help me to develop prototypes (or pilots) test new ideas, and overcome seemingly impossible barriers on the way. The three skills below can used to help solve many problems :

use diagrams & data to quickly
INVESTIGATE, DISCOVER THEORIES & IDENTIFY OPPORTUNITIES
R1, R2, R6 & R17

use diagrams & technical drawings to
DEVELOP CONCEPTS & PROTOTYPES
R3, R4, R5, R7, R8, R11, R18 & R19

use technical drawings to
RECORD ITERATIVE DEVELOPMENT & MOVE TO MANUFACTURE
R9, R14 & R15

But any radical solution would be built on research that would give us a deep understanding of a subject and allow us to **identify root causes**. Understanding the principles (or science) behind why things happen would be useful when solving any new problem or needing to advance knowledge.

the job of a
manager

MANAGING PEOPLE

Some large and expensive assets (ie trains) are designed to last a long time. This means that large maintenance operations (businesses) are required to sustain and continually improve its performance throughout assets life.

The Engineering manager's job is to plan work and resources, manage tasks, and lead teams. Improving quality and reducing costs was key to business success. The tasks include :

develop (**design+build**) & implement (**maintain**) processes

create plans to organise & control resources

leading & developing people

continuously improve the business
& promote best practice

The focus would be on plans, processes, and people rather than developing technical solutions. The first project would be to understand existing processes and develop work planning solutions.

The drive for Continuous improvement and cost-effectiveness would lead to the questioning of how the business worked and how to get the best out of people.

The need for major business change would lead to using the design/build/maintain stages to develop new processes, systems, and organisation structures and to solve other non-engineering problems.

job profile

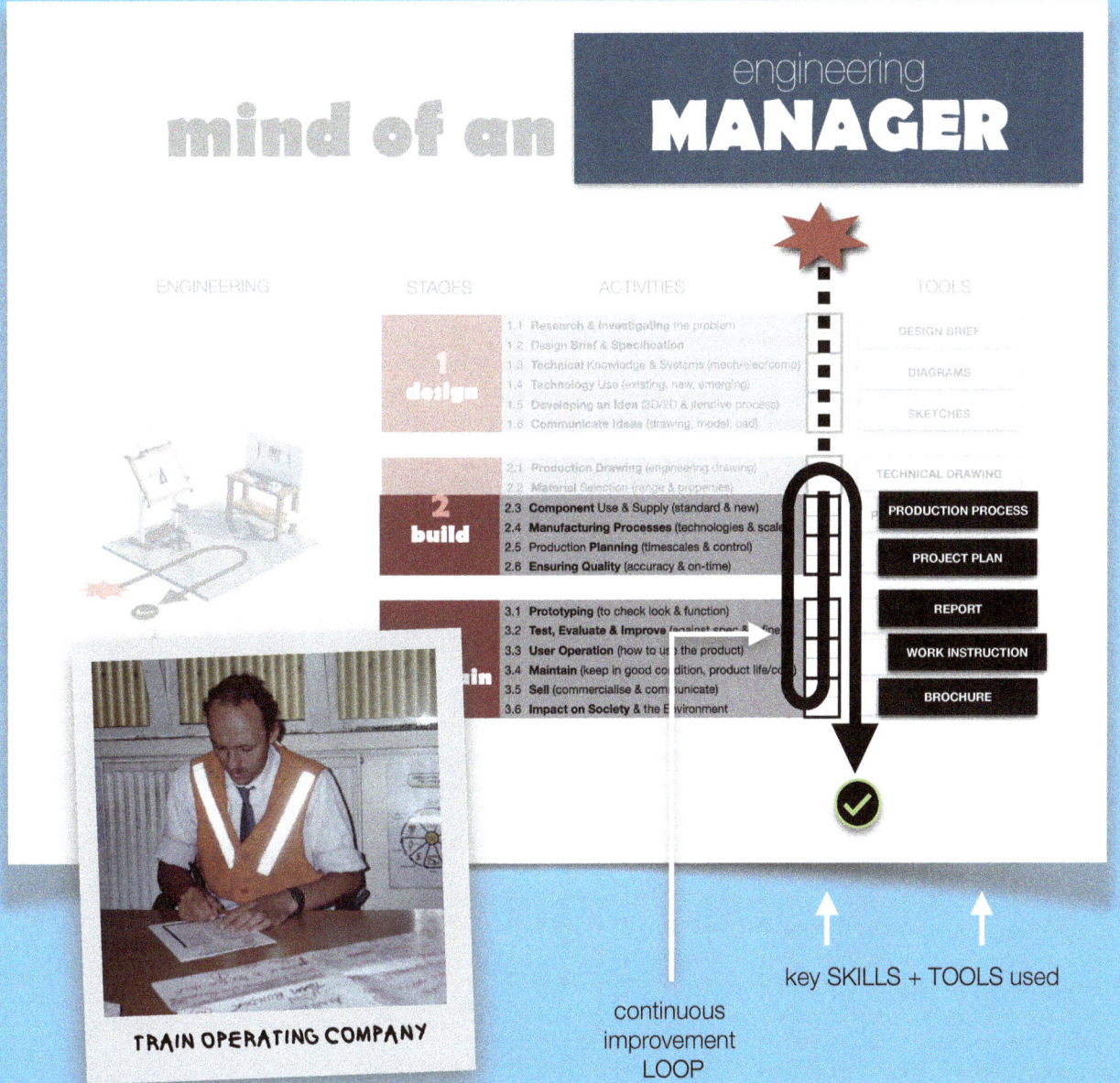

inside the managers office

TRAIN OPERATING COMPANY

A Train Operating Company is a business operating passenger trains on the railway system of Great Britain. The Engineering function is responsible for the performance of the train fleet and the daily servicing, and overhaul of the trains.

MANAGER

Before you can improve performance you need to understand how & why they work now. The constraints of facilities, resources, supply chains, staff quality, etc will help you create realistic improvements.

A new approach to management and employee involvement would enable further improvements. To improve quality and productivity further would require employees to work in teams and be multi-skilled, so new jobs and a new organisation were designed and implemented.

Employee involvement in continuous improvement will be limited by the leadership style. Changing the culture is a long journey & may involve management changes first. Changing processes and/or the culture may require the shape of individual jobs & the organisation to be redesigned.

The engineer has to transition to being a manager and leader of people. Everything is done through people, so you have to learn how to get the best from others.

workplace

TRAIN OPERATING COMPANY

Operated by Intercity then Great Western Railways and operating service from London Paddington to South Wales and the West Country.

ENGINEERING FUNCTION

The Plymouth, Laira is one of several Traction & Rolling Stock Depots carrying out the maintenance and repair of the High Speed Train fleets, as well as various locomotives, coaches, multiple units, and freight vehicles.

ACTIVITIES

Train Servicing & Cleaning, Heavy Maintenance and third party contracts with other train operators.

A LARGE ORGANISATION

A workforce of 400 employees including maintenance, cleaning & movements teams.

technical & business knowledge

TRAIN OPERATIONS

An understanding of how a train company works helps the engineers provide

- The Route (track, signals)
- The Timetable & Services (train types etc)
- The Trains Required (where & when)
- The People (on train crew, station teams, etc)
- Operational Rules including Safety Case

FLEETS OF TRAINS

The engineering function is responsible for the day-to-day maintenance and performance of the train company's fleet of trains. This requires the constant development of solutions to operational failures and more cost-effective maintenance.

The root causes of poor fleet reliability are less likely to be technical and have more to do with maintenance activities and depot or supply chain management.

The diagram opposite shows parts of the engineering business that have an impact on fleet reliability and train service performance This means that multiple systems have to be monitored and managed…..

train operation

KNOWLEDGE OF THE RAILWAY BUSINESS
including the customers, routes, timetable, service needs, train crew, operating rules, service recovery & performance targets

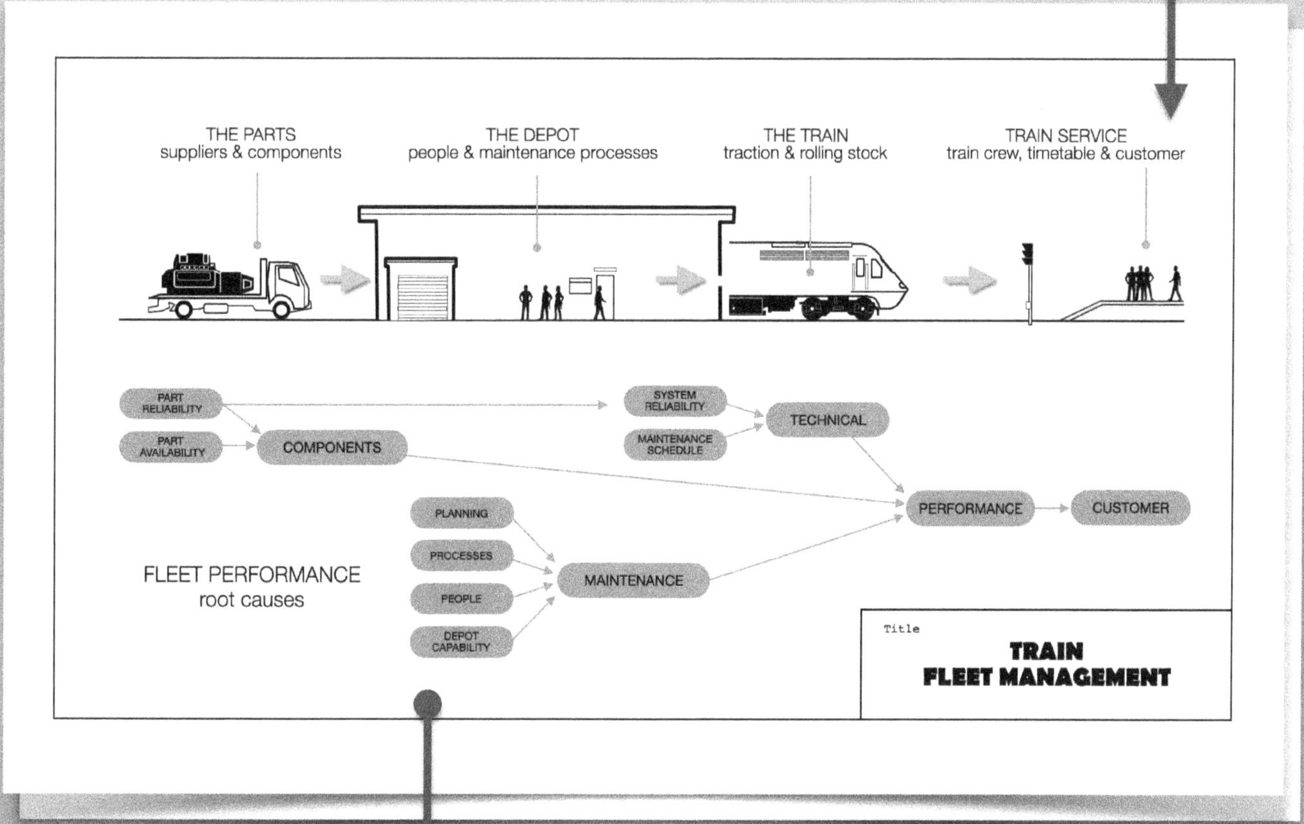

THE PARTS — suppliers & components
THE DEPOT — people & maintenance processes
THE TRAIN — traction & rolling stock
TRAIN SERVICE — train crew, timetable & customer

FLEET PERFORMANCE root causes

- PART RELIABILITY
- PART AVAILABILITY
- COMPONENTS
- SYSTEM RELIABILITY
- MAINTENANCE SCHEDULE
- TECHNICAL
- PLANNING
- PROCESSES
- PEOPLE
- DEPOT CAPABILITY
- MAINTENANCE
- PERFORMANCE
- CUSTOMER

Title: **TRAIN FLEET MANAGEMENT**

BUSINESS MANAGEMENT
continuously reviewing train performance to identify technical, maintenance depot & component improvements

MULTIPLE DEPOTS & FLEETS
different depot locations, sizes & capabilities

engineering management

As a production engineer, I would first need to understand how the business works, its processes, people, and capabilities. I could then manage and continuously improve the performance of the business and train fleets.

understanding the business

BUSINESS KNOWLEDGE

improving train & business performance

PERFORMANCE MANAGEMENT

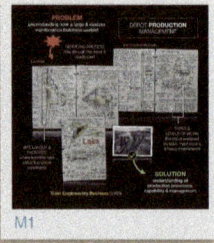

M1

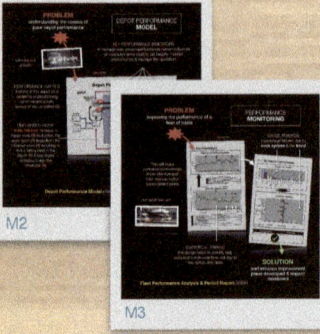

M2

M3

understanding the customers, production activities, processes & management

depot performance model & train performance reports developed to understand causes & areas for improvement

BUSINESS KNOWLEDGE

Understanding the business operations of a train operating company and the engineering function its maintenance depots. This included different customers, contracts and multiple activities. There are two types of maintenance carried out on rolling stock, light maintenance on train operating fleet and heavy maintenance which are commercial (& tendered for) contracts.

BUSINESS PERFORMANCE & MANAGEMENT

Identifying & monitoring key performance indicators to understand how the business is working & identify areas for improvement. Then using this information to more effectively allocate resources to accomplish business goal/targets.

MAINTENANCE activities

IT SYSTEM DESIGN

better management & information

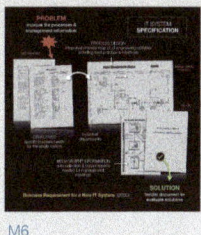

no business case

designing & specifying a new engineering management IT system

winning maintenance contracts

redesign a maintenance processes & planning systems to improve productivity

continuous improvement

TOTAL QUALITY MANAGEMENT

TRAIN CLEANING activities

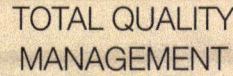

improve standards & productivity

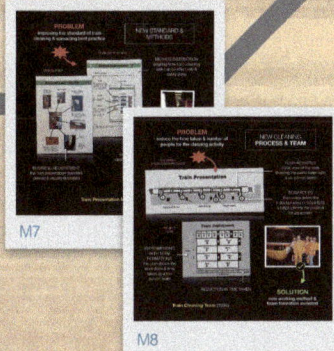

asked 400+ employees if they could identify any business problems & they came up with over 700 issues

created 30+ improvement teams & trained employees in problem solving

developing a new standard, sharing best practice & radical teamwork to improve quality/productivity

---

CONTINUOUS IMPROVEMENT & TQM

Total Quality Management (TQM) is an approach to management that improves the quality of an organisations outputs, through the continual improvement of internal practices. It requires a shift from reactive management to leadership.

change management

I worked as a change manager, solving a whole range of business problems, developing people, and changing the culture. I would use an engineering way of thinking to solve a range of company problems.

ORGANISATION DESIGN & DEVELOPMENT

COMPANY STRATEGY

old organisation structure

improving company performance

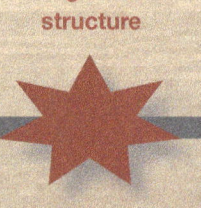

M11

M12

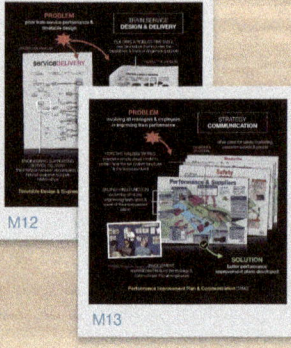

M13

became a Chartered Engineer (CEng) & Member of the Institute of Mechanical Engineers (MIMechE)

designing new job roles, work teams & new organisation structures, then designing & delivering leadership programmes

solving company problems, working cross functionally & helping develop strategies

VISION & STRATEGY

Developing the ability to think about and plan the future. Learning to create an image of what the future could be like & communicate it to others. Then develop a plan to achieve it.

This could include innovative or improved activities, processes, jobs, teams, organisations, culture change etc...

LEADERSHIP & CHANGE

Business success is achieved through people & this requires good, effective leadership. To stay competitive and organisation will have to keep changing. Change management requires leaders to prepare, support, and help individuals, teams and organisations in making organisational change.

AUTHOR & DEVELOPER

spent 15 years researching & creating self help books & business toolkit app, then used with clients to develop engineers

started writing the engineering book & giving talks

creating self help solutions

MORE TRAIN COMPANIES

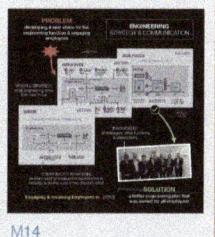

working with directors & 100's of employees to develop new strategies & improvement of company performance

helping others to innovate & change

FREELANCE PROJECTS

BUS COMPANIES

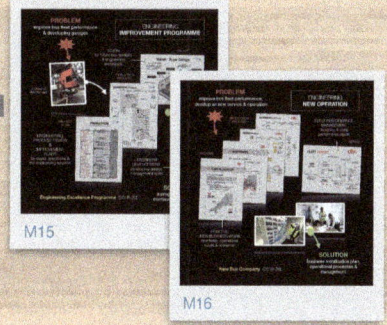

spent 25+ years helping organisations, working in all sectors, with all types and sizes of enterprise

helping to improve & develop bus companies & develop their leaders

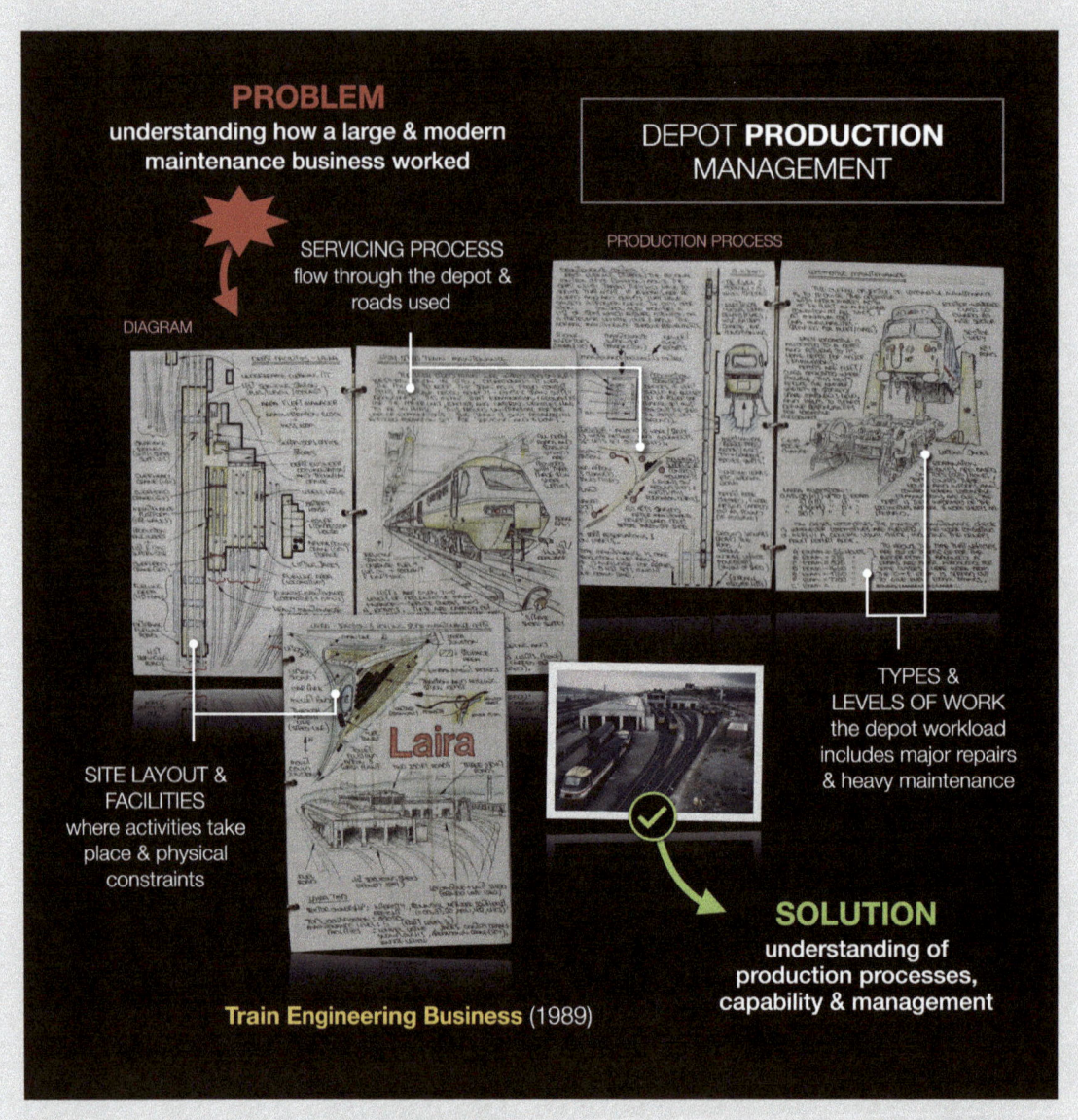

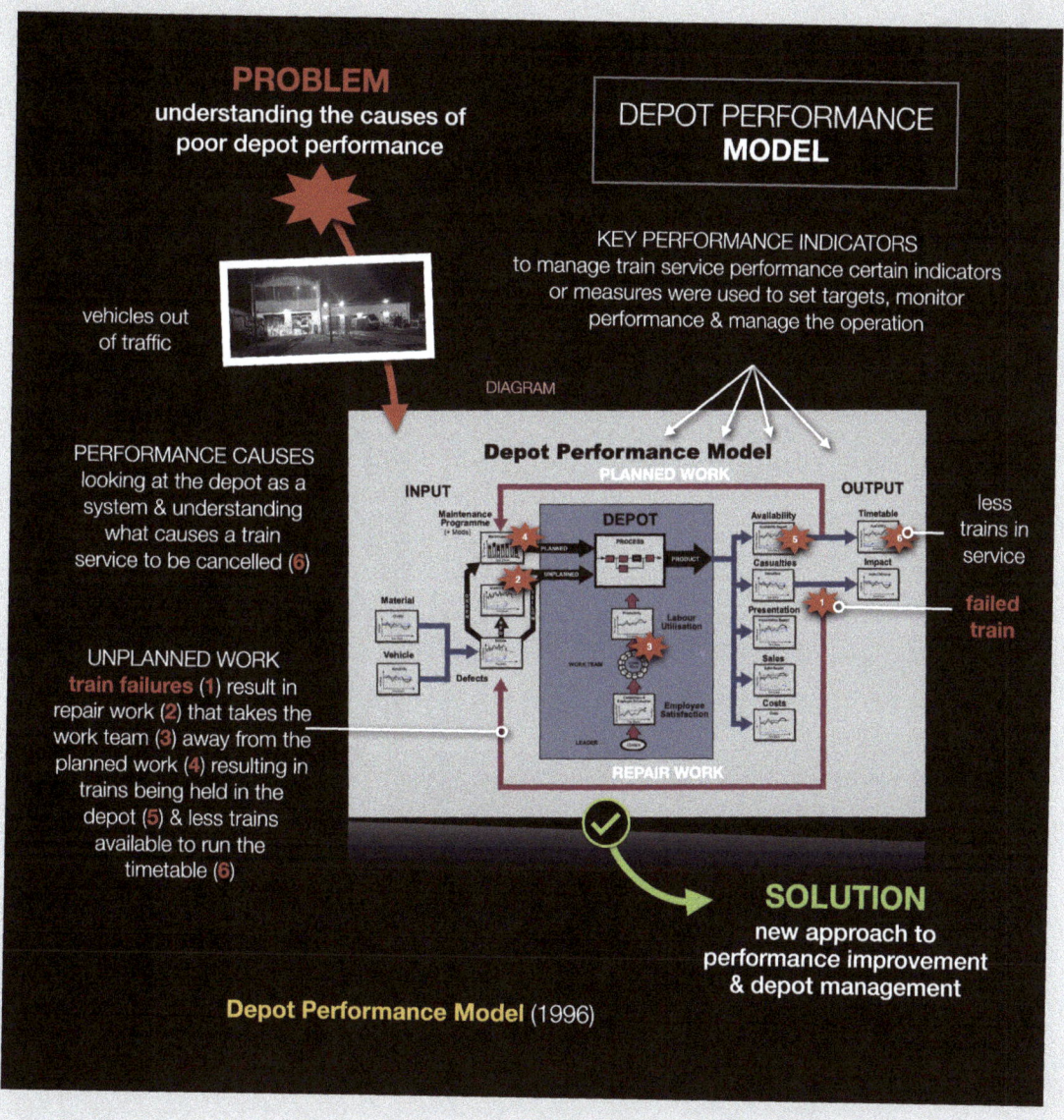

PROBLEM
understanding the causes of poor depot performance

DEPOT PERFORMANCE MODEL

vehicles out of traffic

KEY PERFORMANCE INDICATORS
to manage train service performance certain indicators or measures were used to set targets, monitor performance & manage the operation

PERFORMANCE CAUSES
looking at the depot as a system & understanding what causes a train service to be cancelled (6)

UNPLANNED WORK
train failures (1) result in repair work (2) that takes the work team (3) away from the planned work (4) resulting in trains being held in the depot (5) & less trains available to run the timetable (6)

less trains in service

failed train

SOLUTION
new approach to performance improvement & depot management

Depot Performance Model (1996)

M2 PERFORMANCE : MODEL

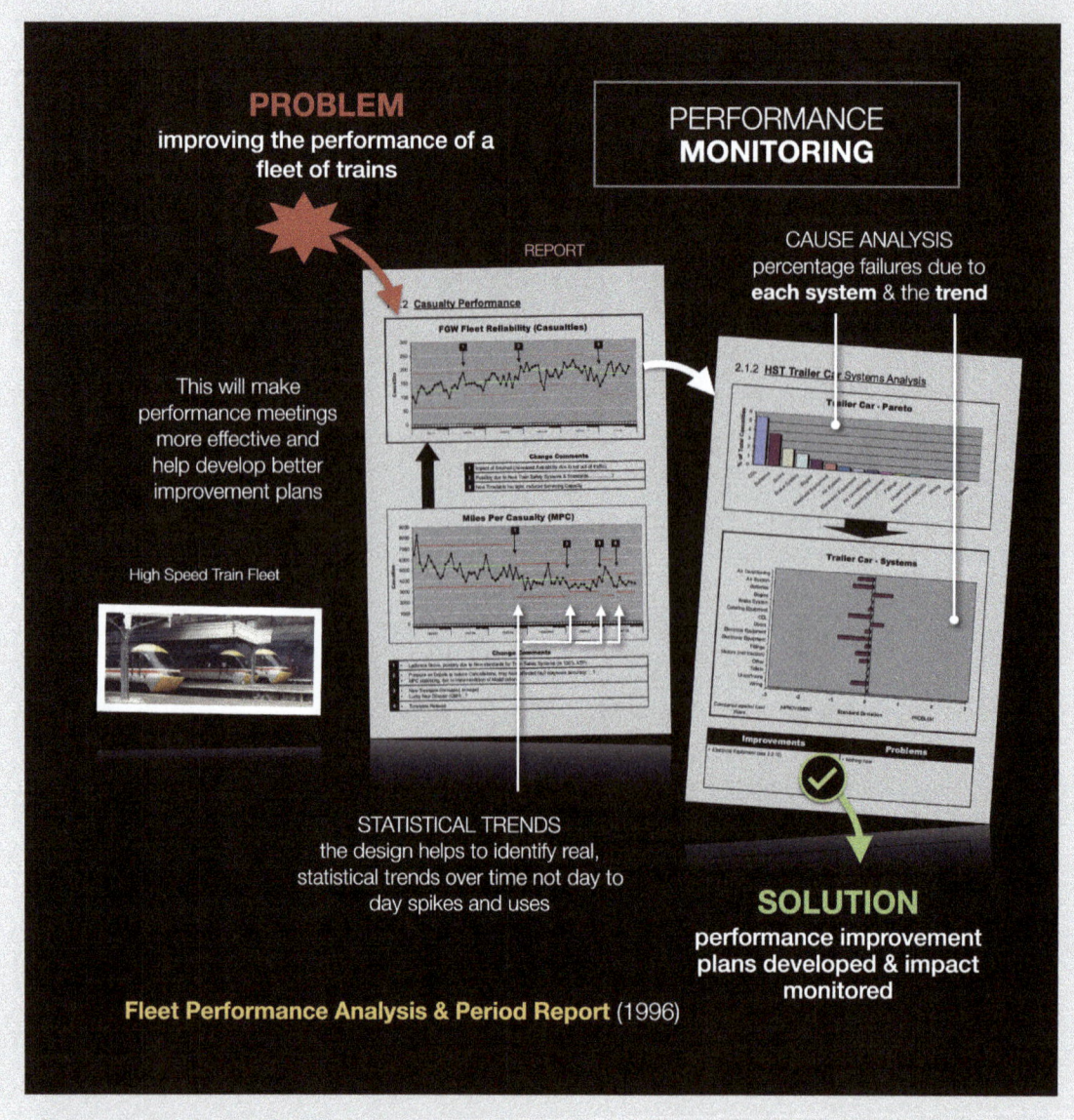

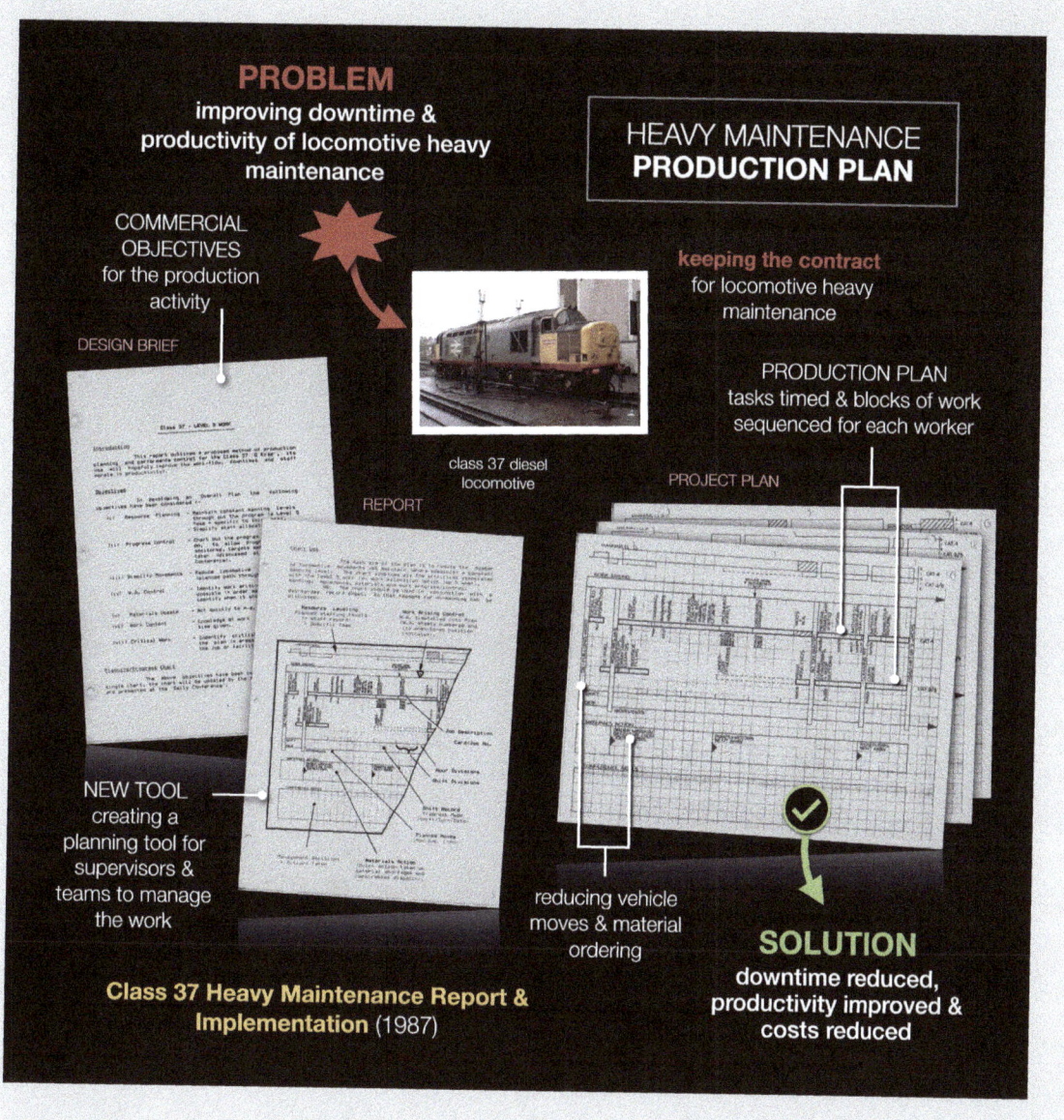

MAINTENANCE PROCESSES

M5

High Speed Train Heavy Maintenance (1987)

PROBLEM
improve the heavy maintenance on high speed train power cars

DESIGN BRIEF

PRODUCT MAINTENANCE

PRODUCTION PROCESS & PLANNING TOOL
most effective & logical sequencing of activities, starting at the top & working down

PRODUCTION PROCESS

WHO DOES
responsibility & time needed for each activity

BLOCKS OF WORK
maintenance schedule tasks grouped into lumps of work, including moving the vehicle

TRACK PROGRESS
work done recorded to show progress & estimate release date

SOLUTION
new plan & way of working implemented that reduced time to do the work

high speed train power cars

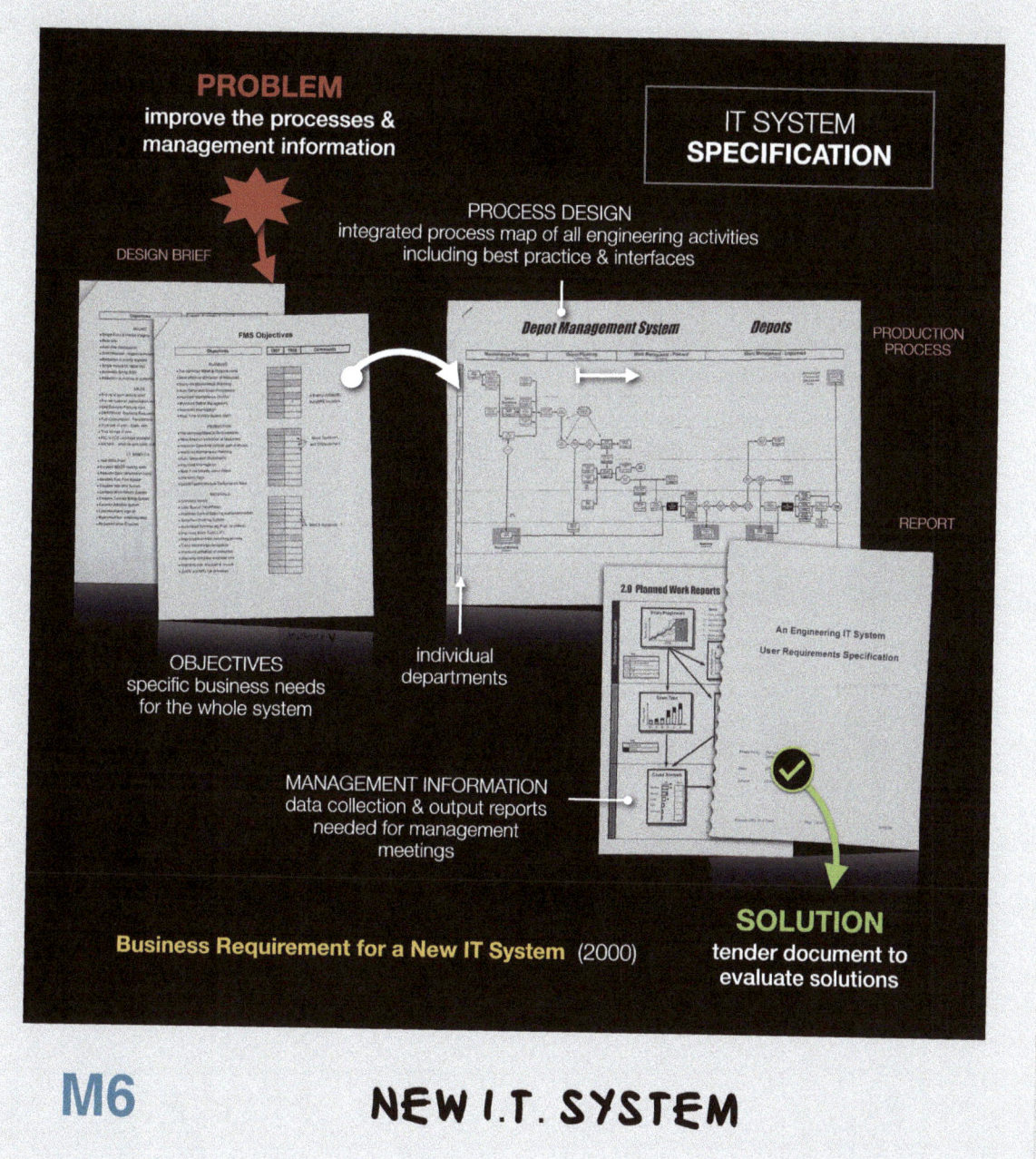

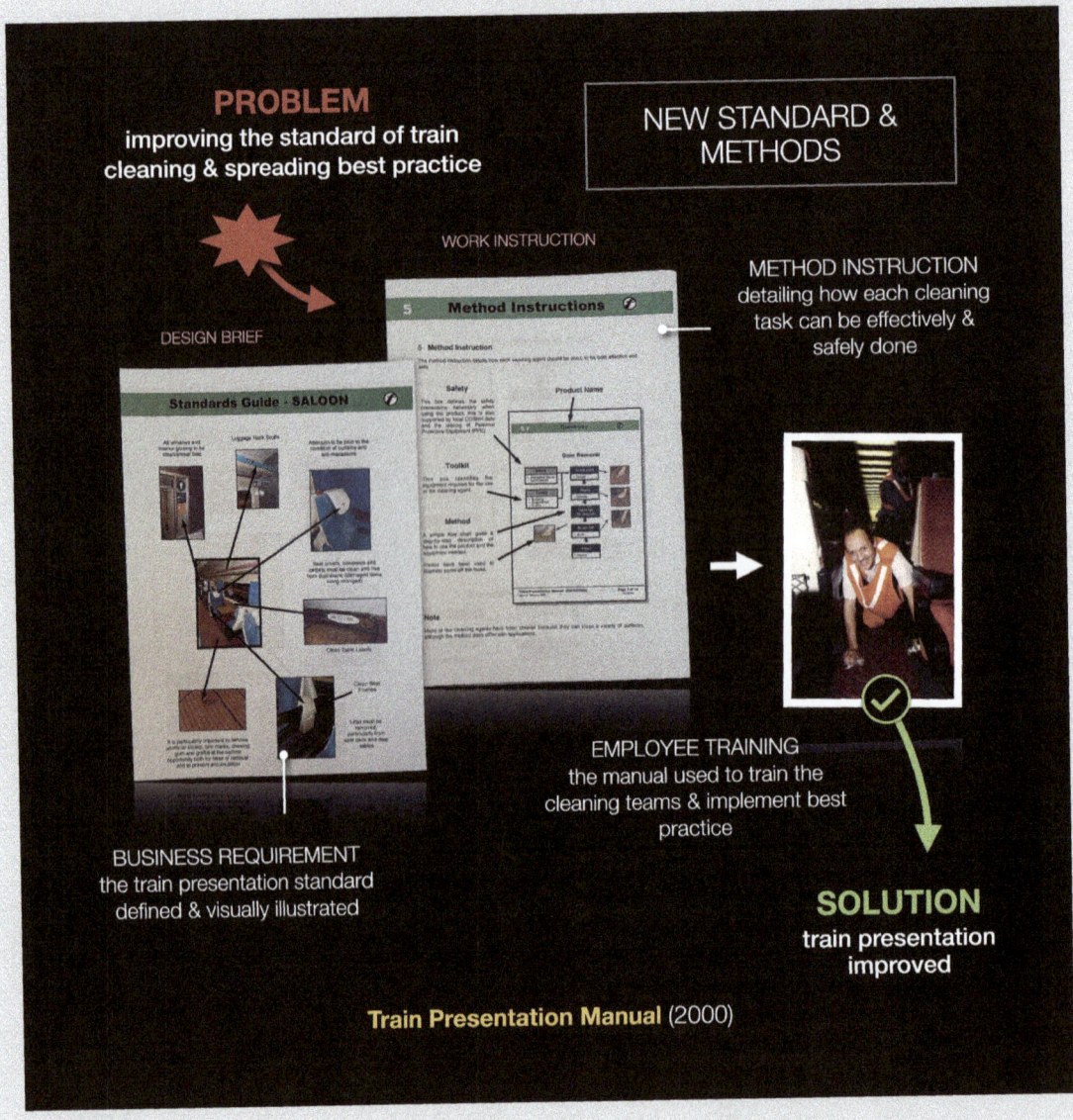

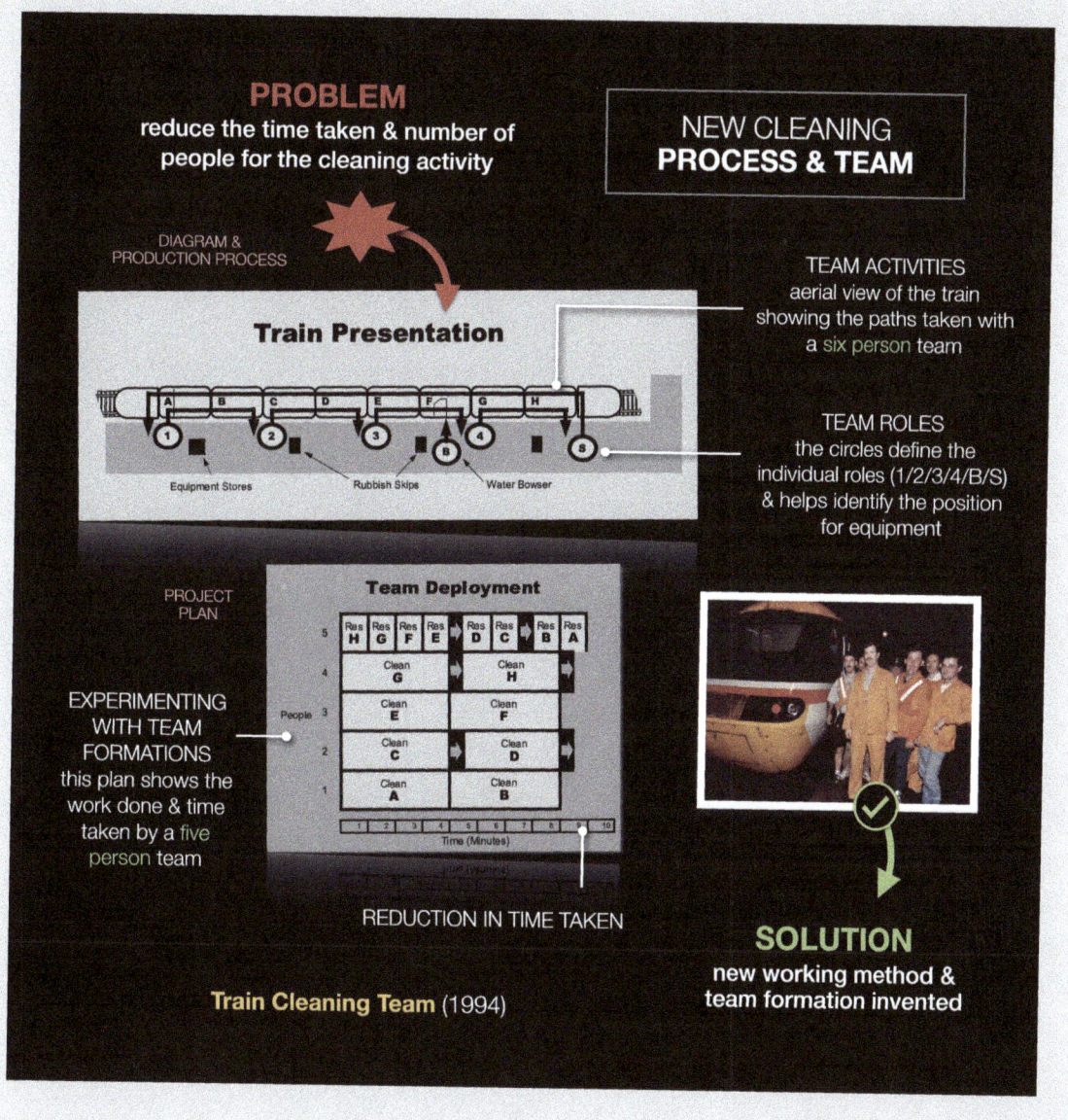

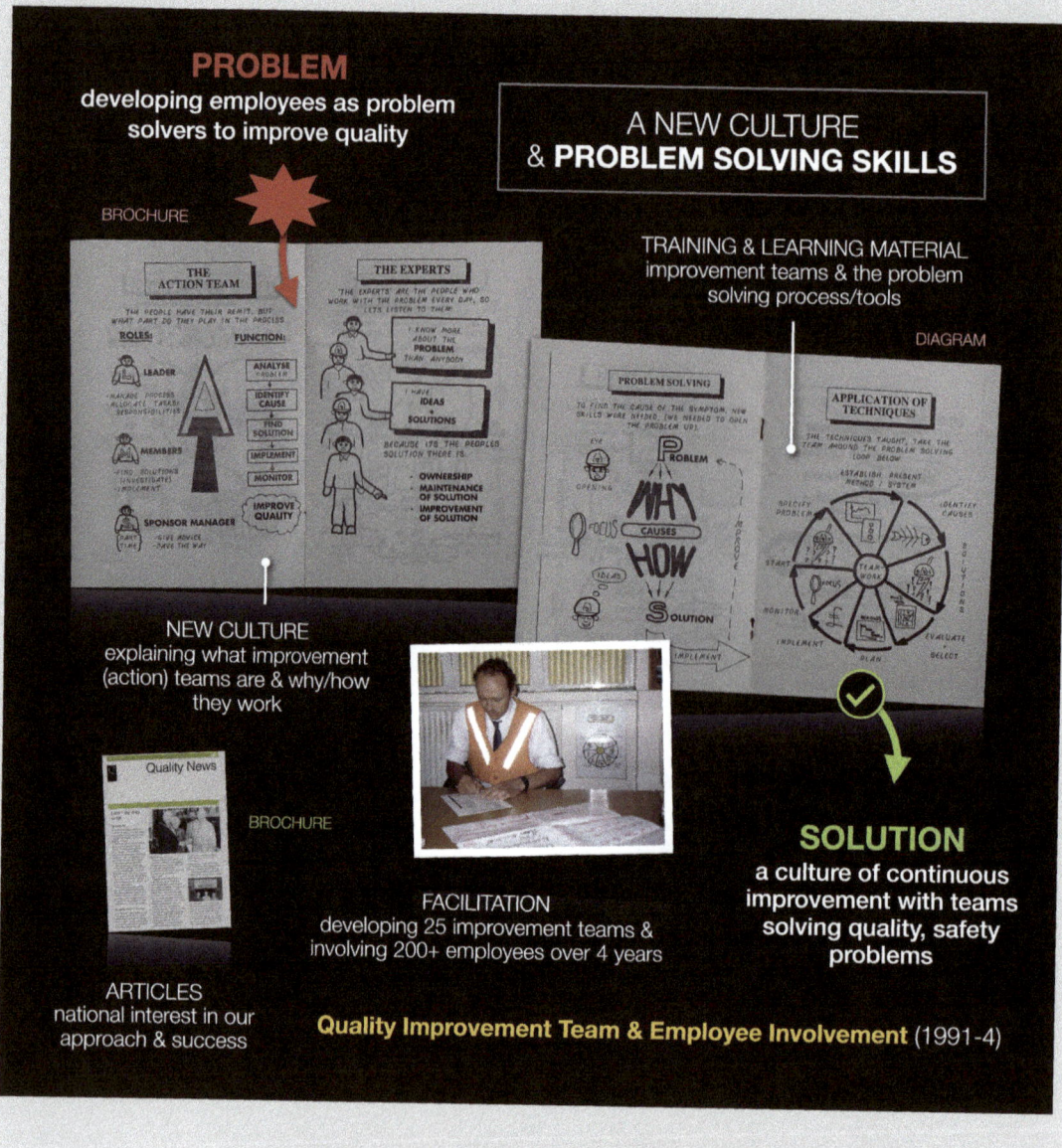

M9 — EMPLOYEE INVOLVEMENT

A NEW CULTURE & PROBLEM SOLVING SKILLS

PROBLEM
developing employees as problem solvers to improve quality

NEW CULTURE
explaining what improvement (action) teams are & why/how they work
BROCHURE

TRAINING & LEARNING MATERIAL
improvement teams & the problem solving process/tools
DIAGRAM

FACILITATION
developing 25 improvement teams & involving 200+ employees over 4 years

ARTICLES
national interest in our approach & success

SOLUTION
a culture of continuous improvement with teams solving quality, safety problems

Quality Improvement Team & Employee Involvement (1991-4)

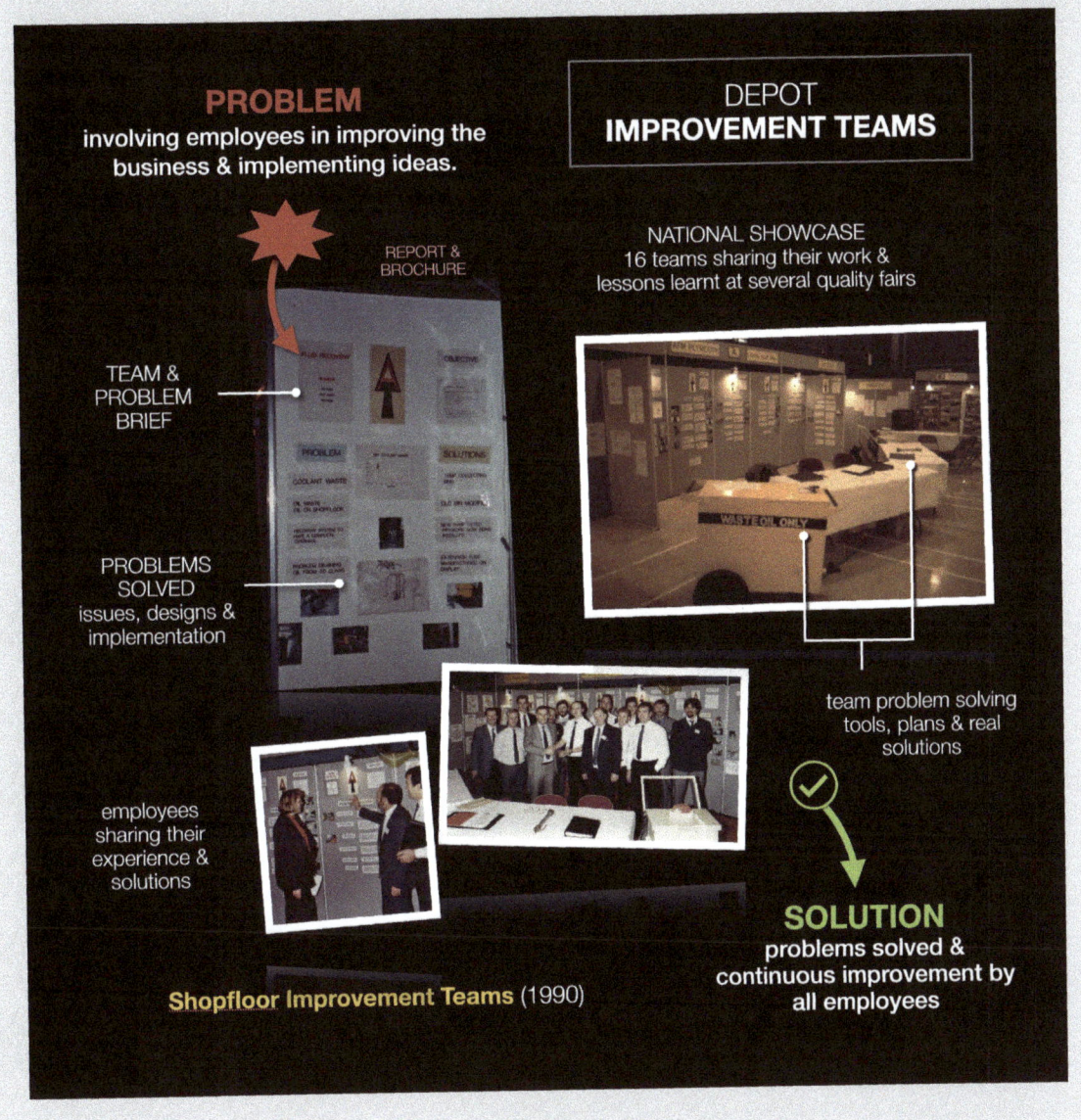

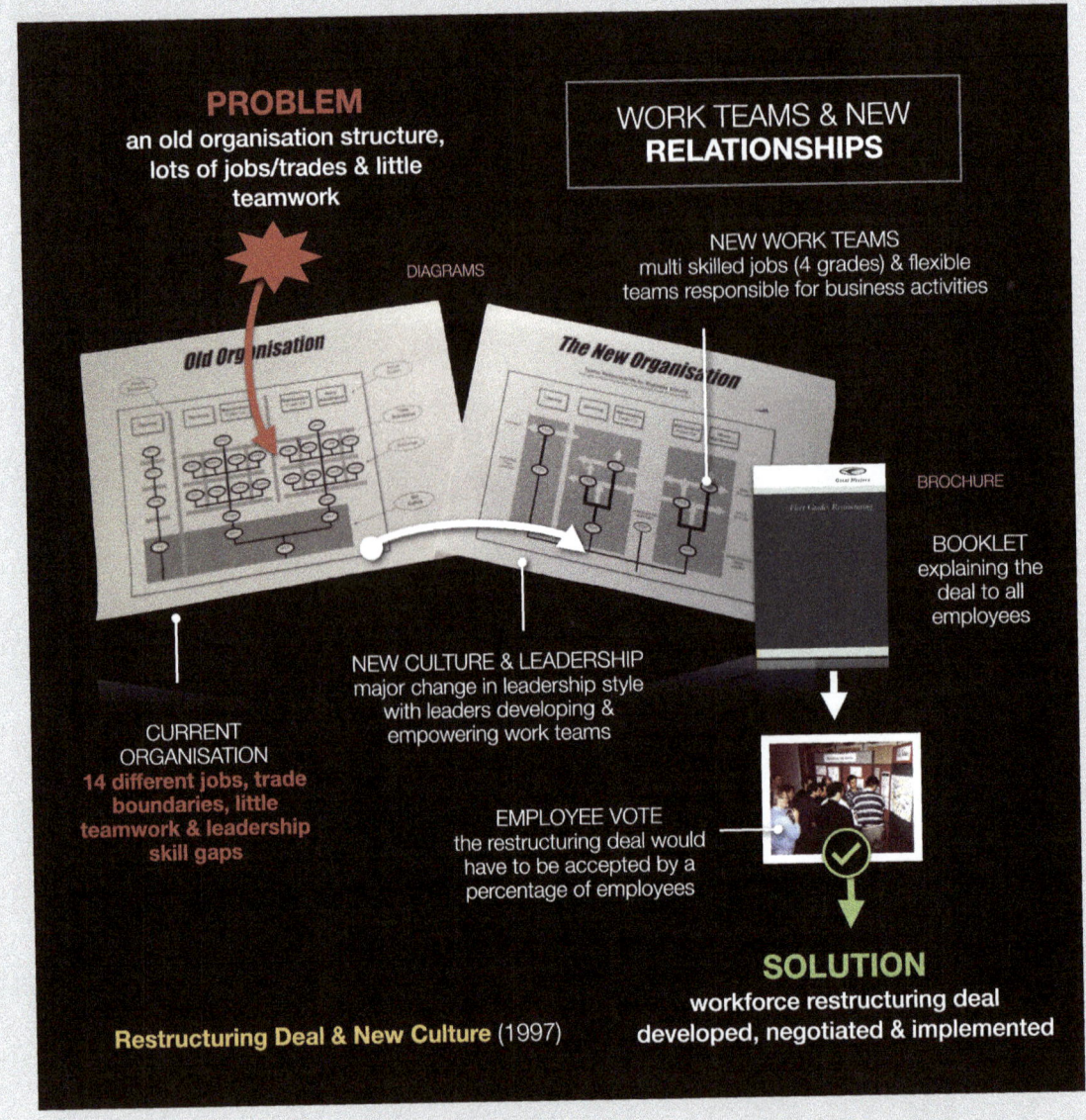

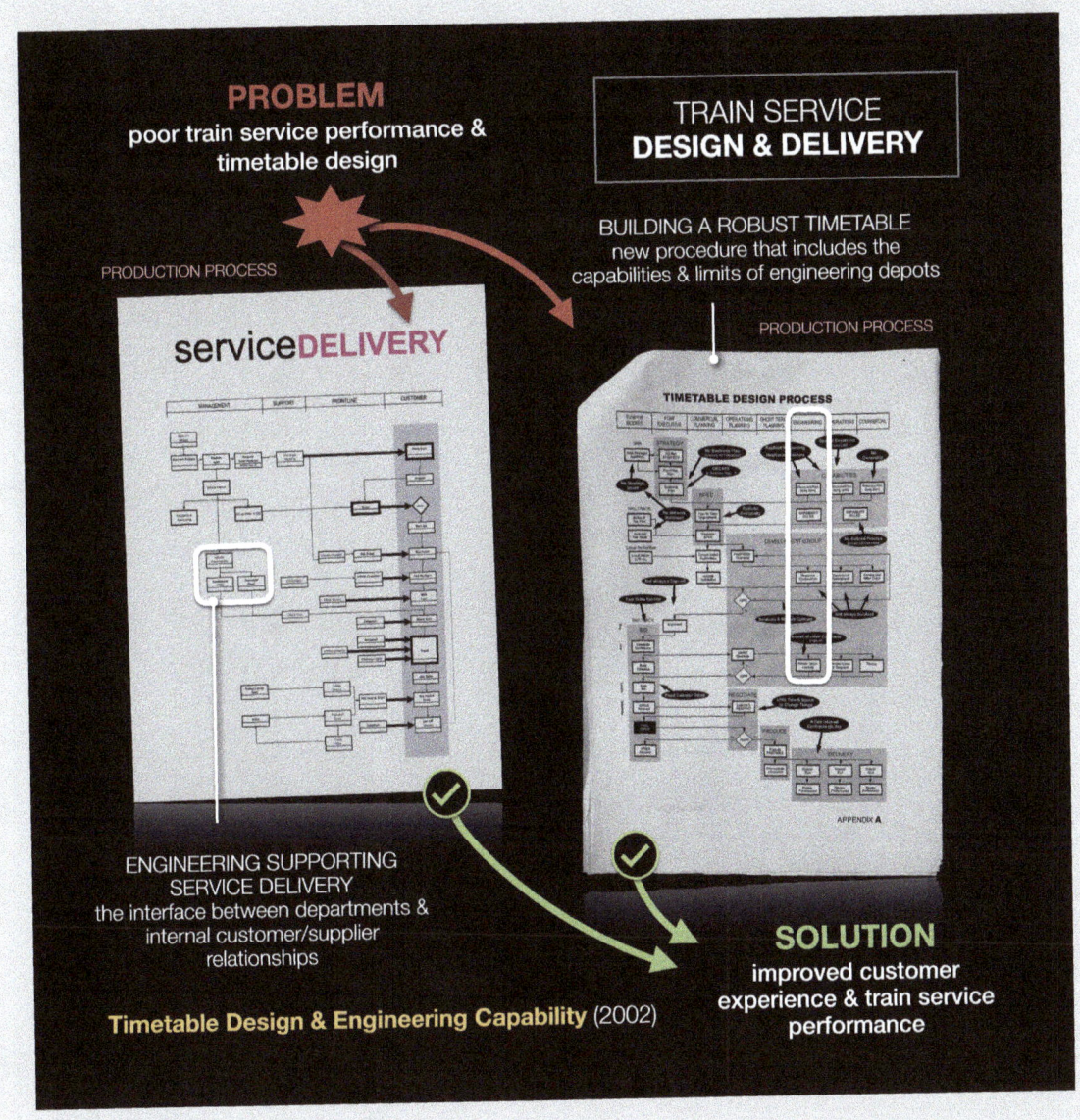

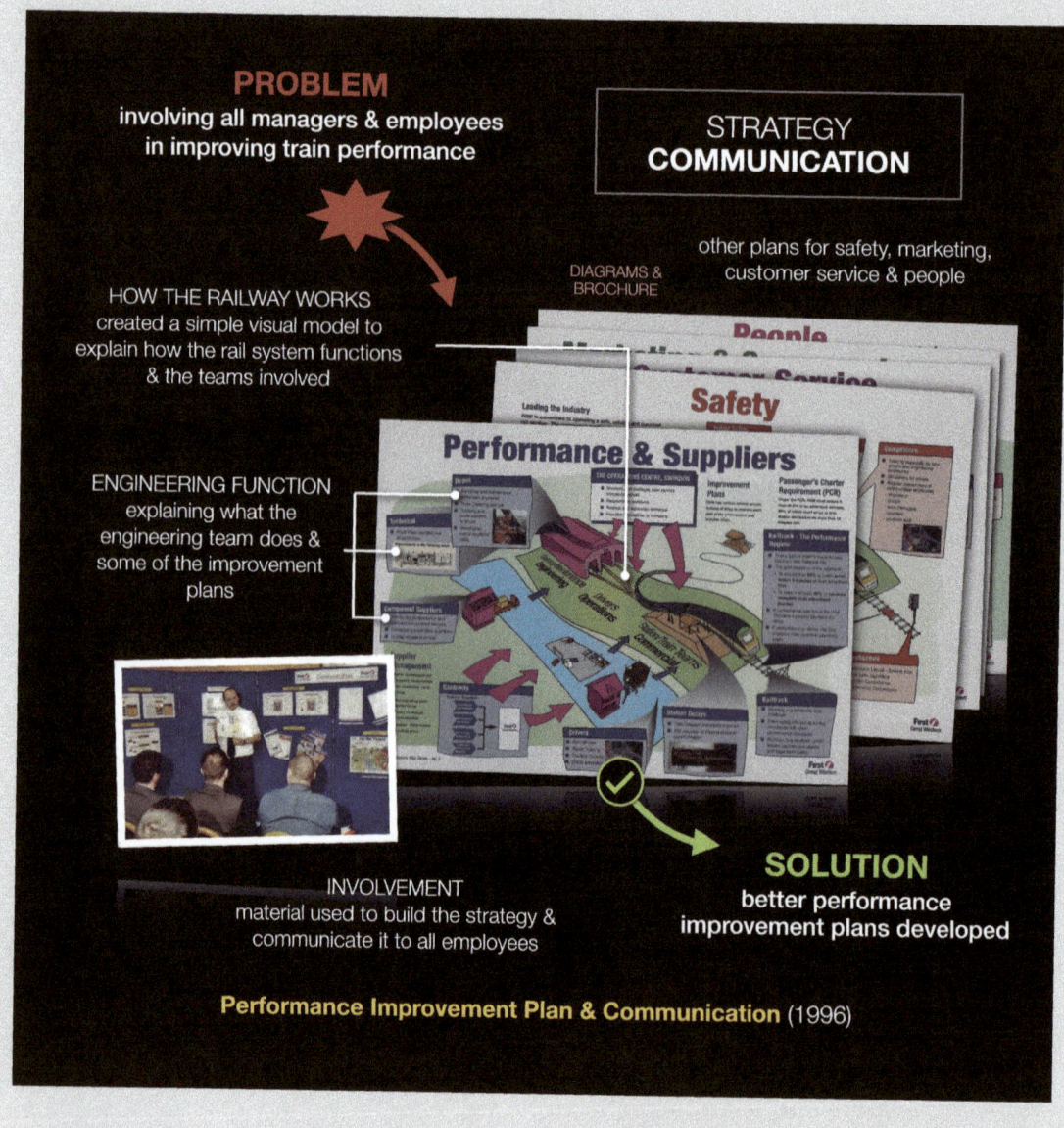

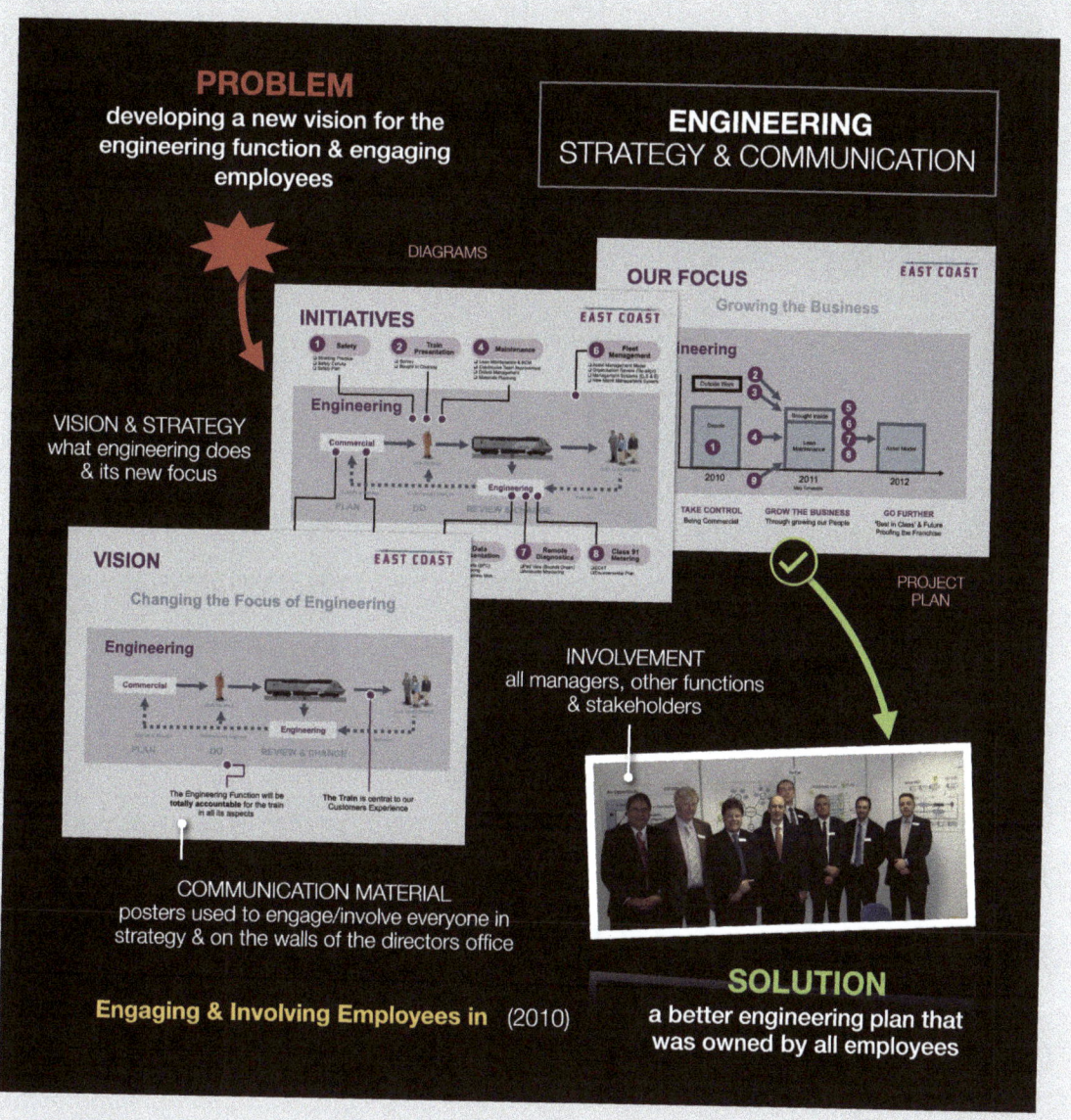

M14 — ENGINEERING STRATEGY

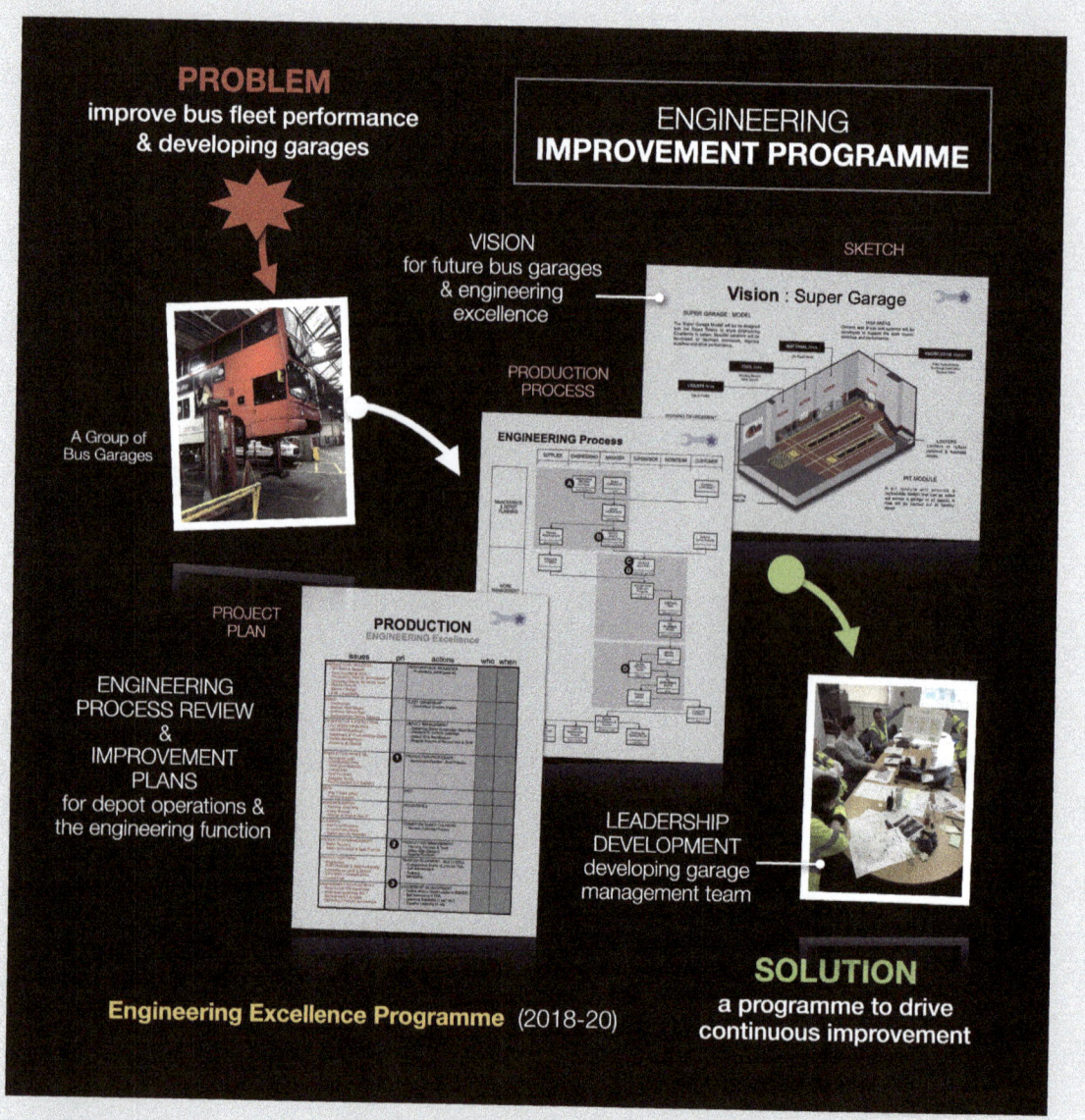

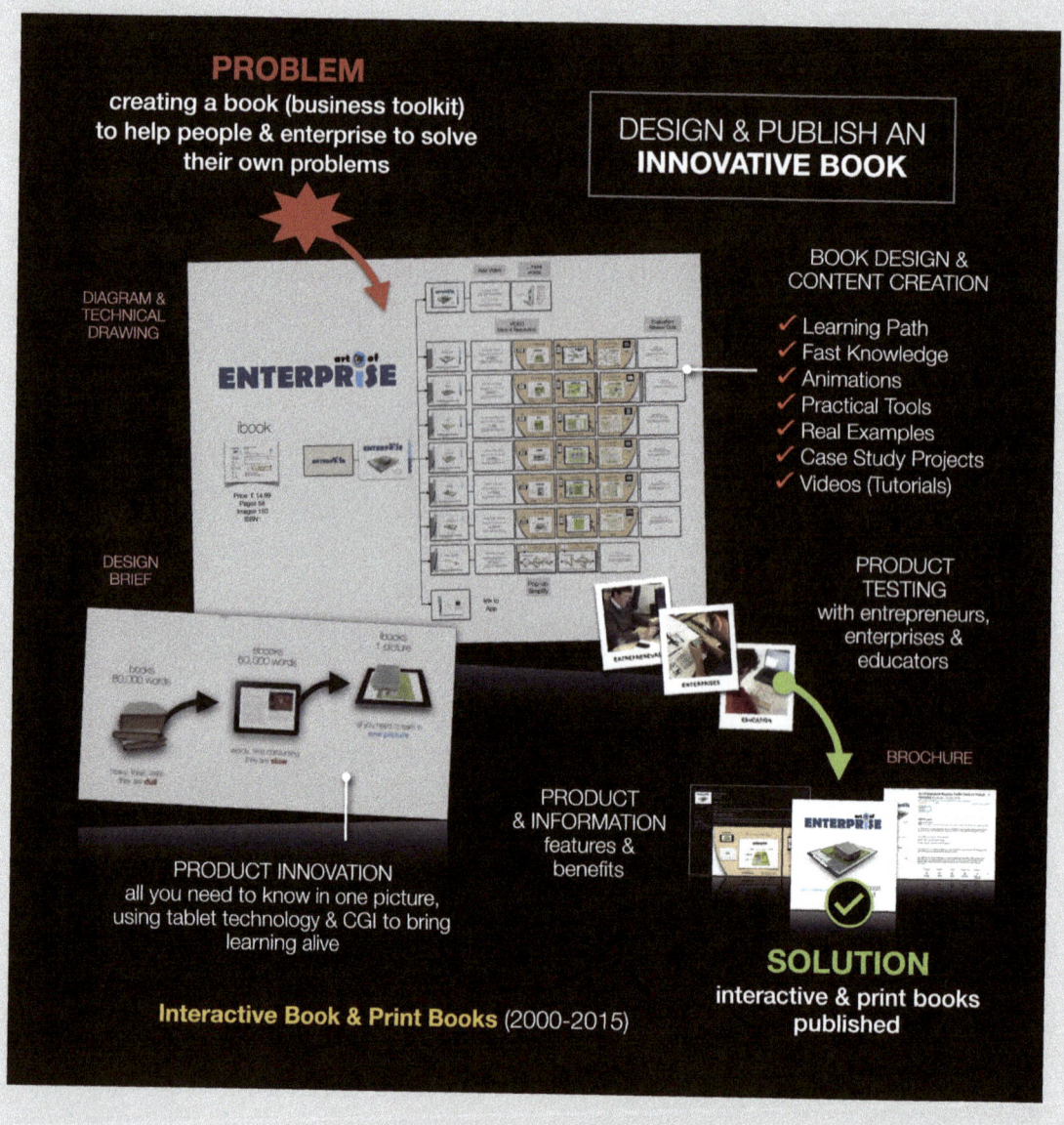

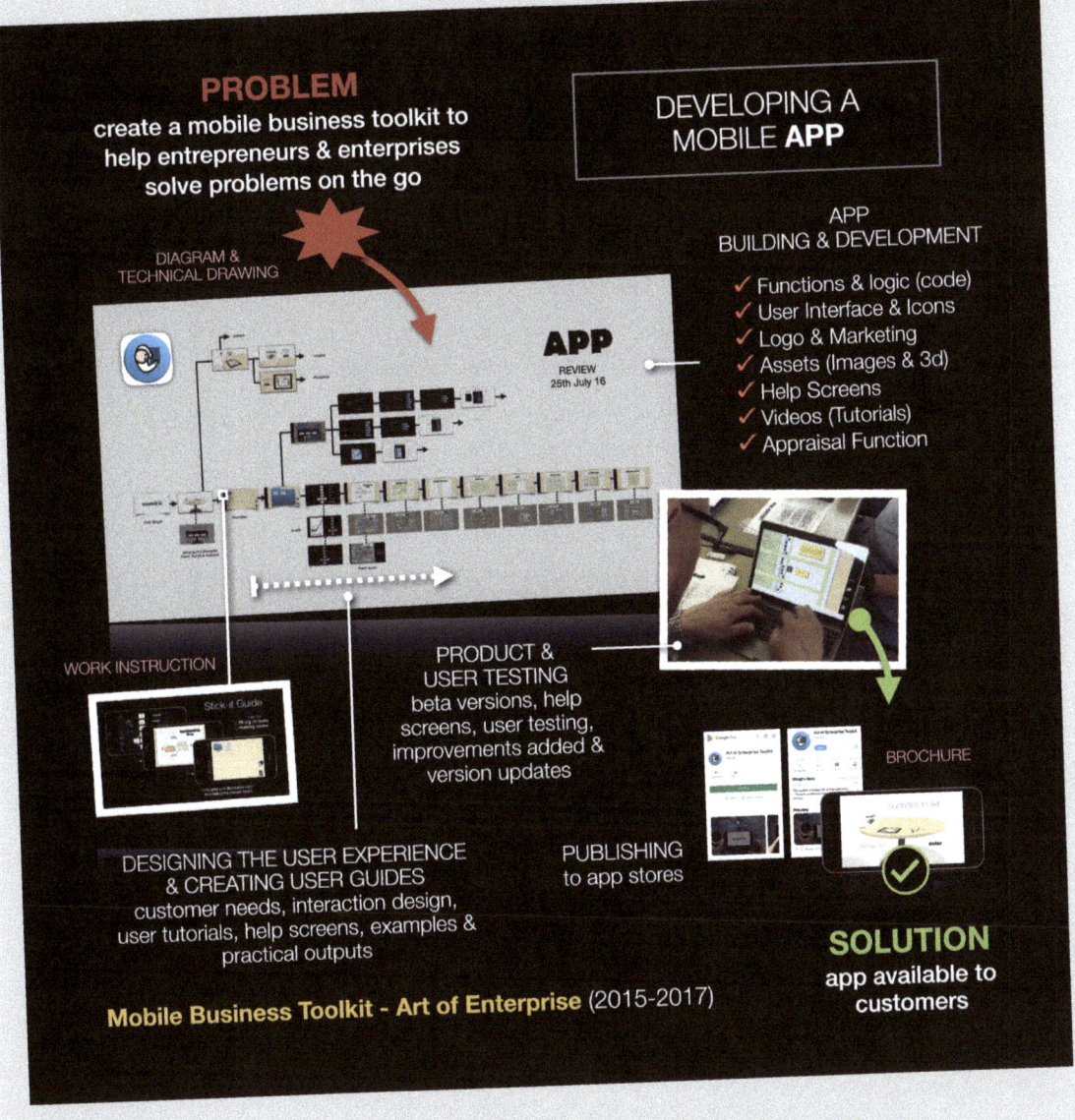

INSIGHTS on management

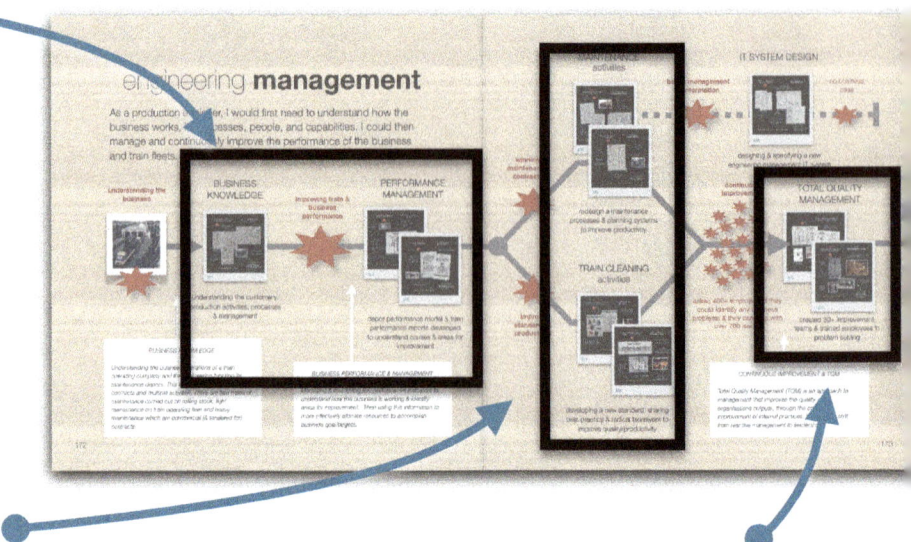

KNOW HOW THE BUSINESS WORKS
understanding how enterprises work, their processes, organisation & performance, to understand why things work or have been designed that way

USE PROCESSES + PLANS TO MANAGE & IMPROVE ACTIVITIES
develop plans to organise & control activities & resources, review performance to identify areas for improvement

INVOLVE EMPLOYEES IN THE IMPROVEMENT
involve employees in problem solving, because they normally know more about the problems & have ideas & solution ownership

---

MANAGEMENT TIPS for Continuous Improvement

- ☑ first, understand **how the business works**
- ☑ use processes & plans to **manage** activities/resources
- ☑ review performance & **involve employees in improvements**

leadership & change

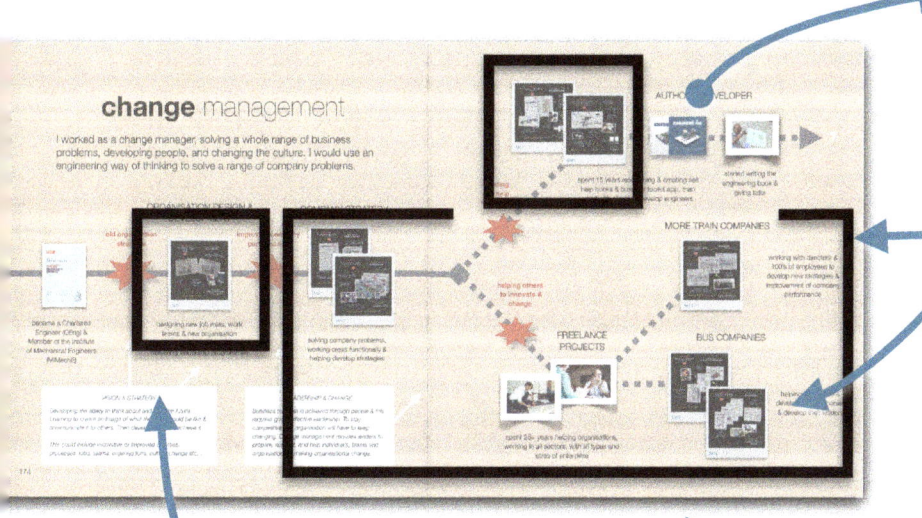

CREATING OTHER INNOVATIVE PRODUCTS
using the engineering approach to innovation & design to create books, apps & learning resources

DEVELOPING PEOPLE & LEADERS
leading teams & developing individuals through management & mentoring

CHANGE MANAGEMENT
employee engagement, leading teams & working with others

CREATE A VISION & STRATEGY THEN COMMUNICATE IT
explain where things are now, what the future will look like & then how to get there

LEADERSHIP TIPS for Making Changes

- ☑ create **a vision** for the future & strategy to get there
- ☑ engage employees & teams through **the change**
- ☑ continuously **developing people** & the organisation

transferable skills

CREATING CHANGE

Most business or organisation problems can be solved using 'an engineering approach' to understand how enterprises work, their processes, organisation & performance, to understand why things work or have been designed that way. The three skills below can used to help solve many problems :

use diagrams & data to
UNDERSTAND HOW A BUSINESS WORKS & PERFORMS
M1, M2 & M3

use diagrams & drawings to
CREATE A VISION & DESIGN THE FUTURE
M11, M12, M13, M14, M15 & M16

use diagrams & drawings to
COMMUNICATE & IMPLEMENT CHANGE & DEVELOP PEOPLE
M11, M13, M14

But create a vision for the future & strategy to get there. engage employees & teams through the change. continuously developing people & the organisation.

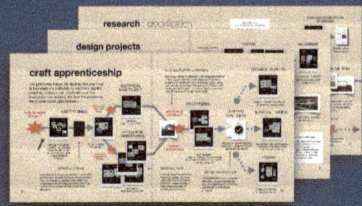

how engineers
DEVELOP

realities of **learning & development**

The motivation to grow, the desire to do your best, comes from us or sometimes it's because we have to stay alive and get a job. Responsibility for the learning is ours, we have to take ownership of it, determine what needs to be developed and create a plan. Real-life learning is driven by us, we're seldom spoon-fed.

Learning opportunities are all around, it's down to you to spot and grab them. We grow by trying new things, things that we aren't quite capable of doing, and the stretch, of going out of our comfort zone. The most effective way is to learn from others who have dealt with these problems before.

development is **through projects** and **problem-based** learning

Engineering has to be learned by doing rather than theorising. Practice, hands on experience, and on the job are generally the means of acquiring knowledge. Learning by doing real work, solving real problems, and understanding the relevance of knowledge and skills.

Recording and reflecting on what we have learned completes the learning cycle, helps us consolidate what we have learned and gives us a foundation to build on. Logbooks became my way of reflecting on my learning journey, track my progress, and get my next job… leading to an unusual career path.

how engineers learn & develop

LEARNING BY SOLVING
REAL PROBLEMS

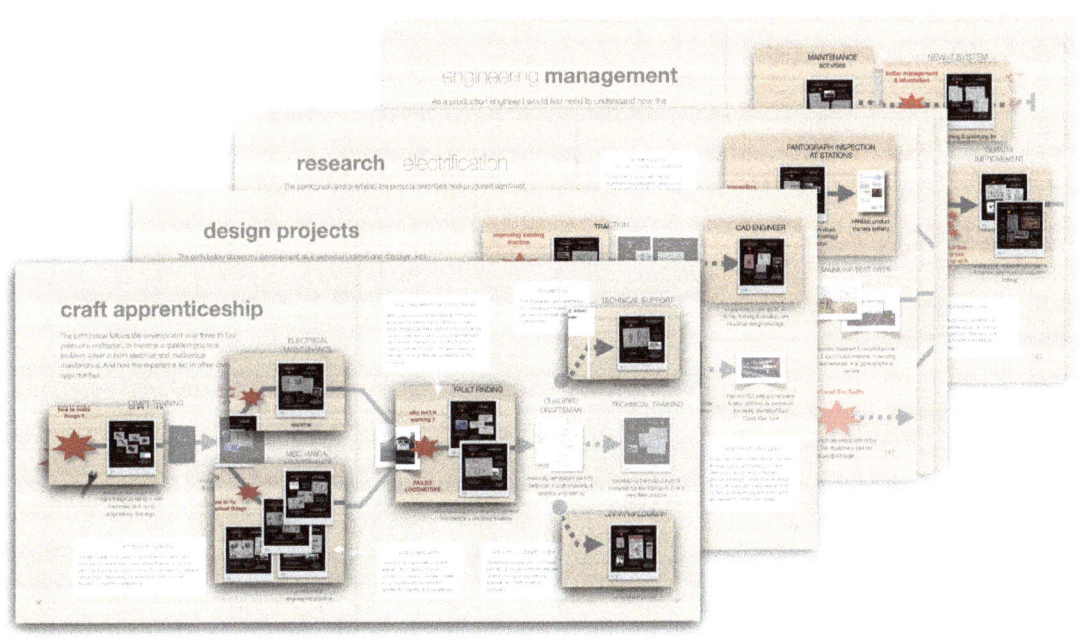

PLAN & DO
ACTIVITY

RECORD & REVIEW
ACTIVITY

continuous
development

Professionals have an obligation to maintain their competence through continuous professional development (CPD), and to support the learning of others. Recording and understanding professional development is sometimes difficult. To help for each job I have mapped out the development path I took, showing the problems I had to solve and what that led to. The journeys are different for each job, for the craftsman it is the defined path of an apprenticeship, for the other jobs my development is informal learning through the challenges and opportunities of working life.

Reflective practice not only helps us accumulate knowledge and experience but also gives a sense of achievement and builds personal confidence. Records are not only evidence of competence, they also help us figure out patterns and see the personal strengths and potential. It's important to have a mentor and seek feedback, because other people see your potential.

personal **strengths** & potential

There are many paths to a career in engineering, and the choice of course should suit your interests. It's beginning to find out what we don't know and what we need to know, then choosing the most effective development activity. It is important to keep in mind that we must not only demonstrate competence, but also show our talent and sell ourselves if we are going to be able to get the next job.

It is sometimes difficult to record your professional development, but it is necessary for you to maintain and improve your skills in order to keep up with the pace of change. In order to ensure that you remain competent, CPD is the maintenance and development of knowledge and skills, and is key for continued professional success

personal potential & progression

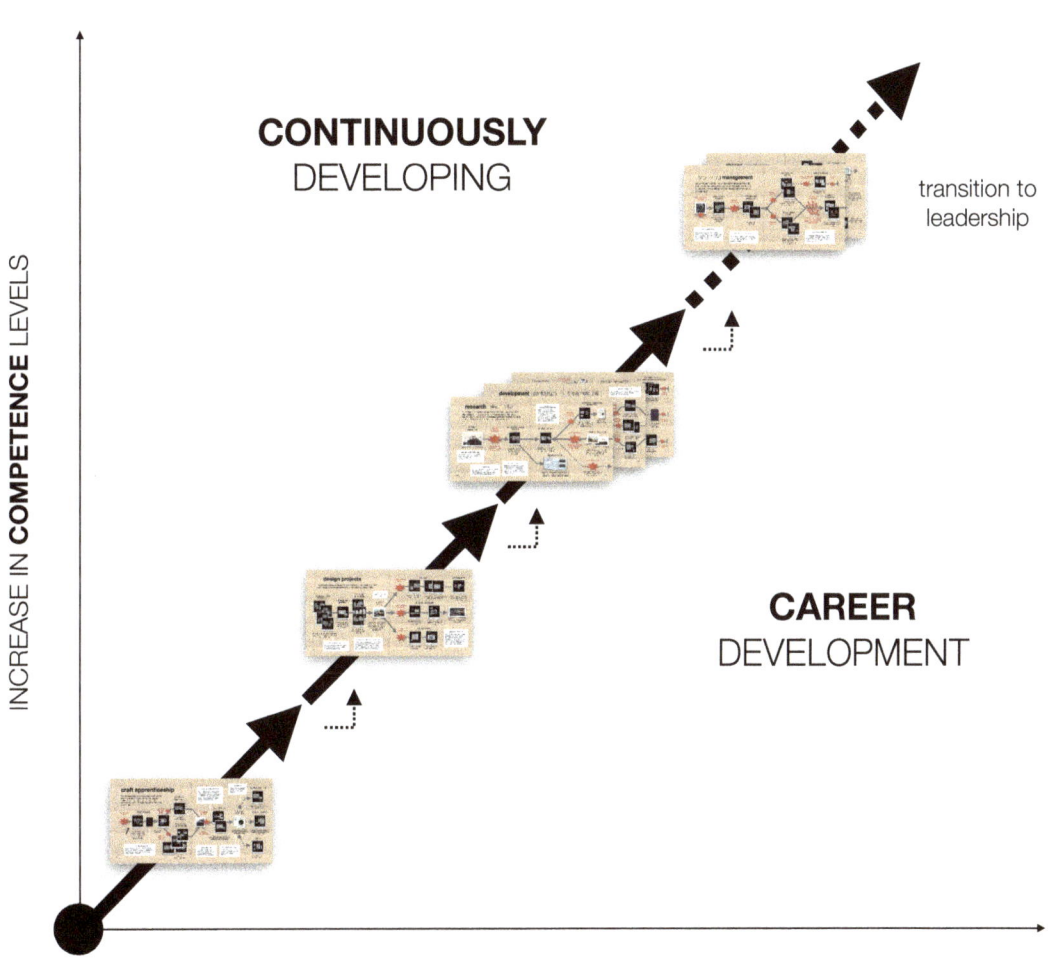

technical **knowledge**

Engineering is a broad field with many different disciplines. The four traditional disciplines of engineering are civil engineering, mechanical engineering, electrical engineering, and chemical engineering, with numerous other engineering sub-divisions. The majority of engineers specialise in one or more fields, but they often work on projects with inter-disciplinary teams.

To ensure and assure that a person can do a job or task, engineers have to become qualified. Relevant qualifications are determined by the engineering requirements of a job and its level. These are the minimal requirements laid down by professional bodies responsible for regulating this field.

> Engineering is the application of **science**, **mathematics**, and technology to solve technical problems

Engineering is a highly scientific and technical profession, and engineering science and methods of mathematical analysis are the starting point of any new work. Scientific, engineering, and mathematical principles provide guidance in creating a product/system that is safe and functional. Allowing engineers to carry out detailed calculations of forces and deflections, contraptions and flows, voltages and currents, that are required to test a proposed design on paper.

The foundation of an engineering education are basic sciences and mathematics, however, they also require training in engineering sciences such as structural and fluid mechanics, thermodynamics, and electrical science. More complex problems require a deeper theoretical understanding and increasingly complex mathematics to calculate, and predict the performance of a solution.

professional & technical education

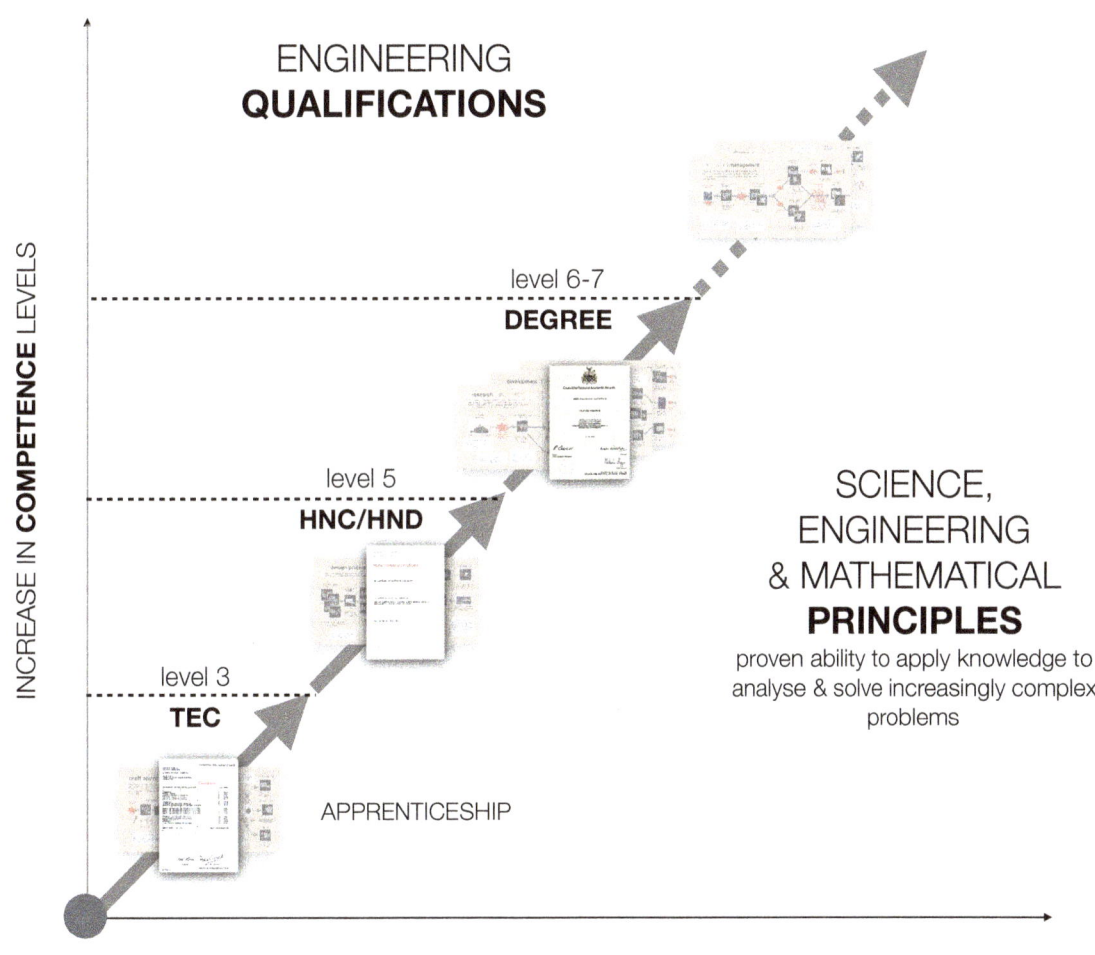

professional standards

The engineering profession has established a set of competence standards for the different levels, that every engineer has to achieve and demonstrate. The grades of professional engineers have been defined as engineering technician (EngTech), incorporated engineer (IEng), and chartered engineer (CEng), these are internationally recognised titles.

Professional registration verifies that an individual has reached a set standard of knowledge, understanding, and occupational competence. It means demonstrating competence by providing evidence of experience. However, meeting the requirement requires the necessary foundation of underpinning logic and analytical capabilities.

achieving and demonstrating **professional competence**

The standards are used to assess the competence and commitment of individual engineers and technicians in the UK engineering profession. A descriptor for each title is shown opposite to show the types of problem and problem solving used at each level. To keep up with change, keep up to date with technology, and stay at the cutting edge we need to maintain our competence.

The standard defines the required ability to carry out engineering tasks successfully and helps us identify areas where we lack the appropriate competence. The jobs and work examples included in this book help you understand the breadth of experience required to achieve the professional standard and become registered.

professional recognition & titles

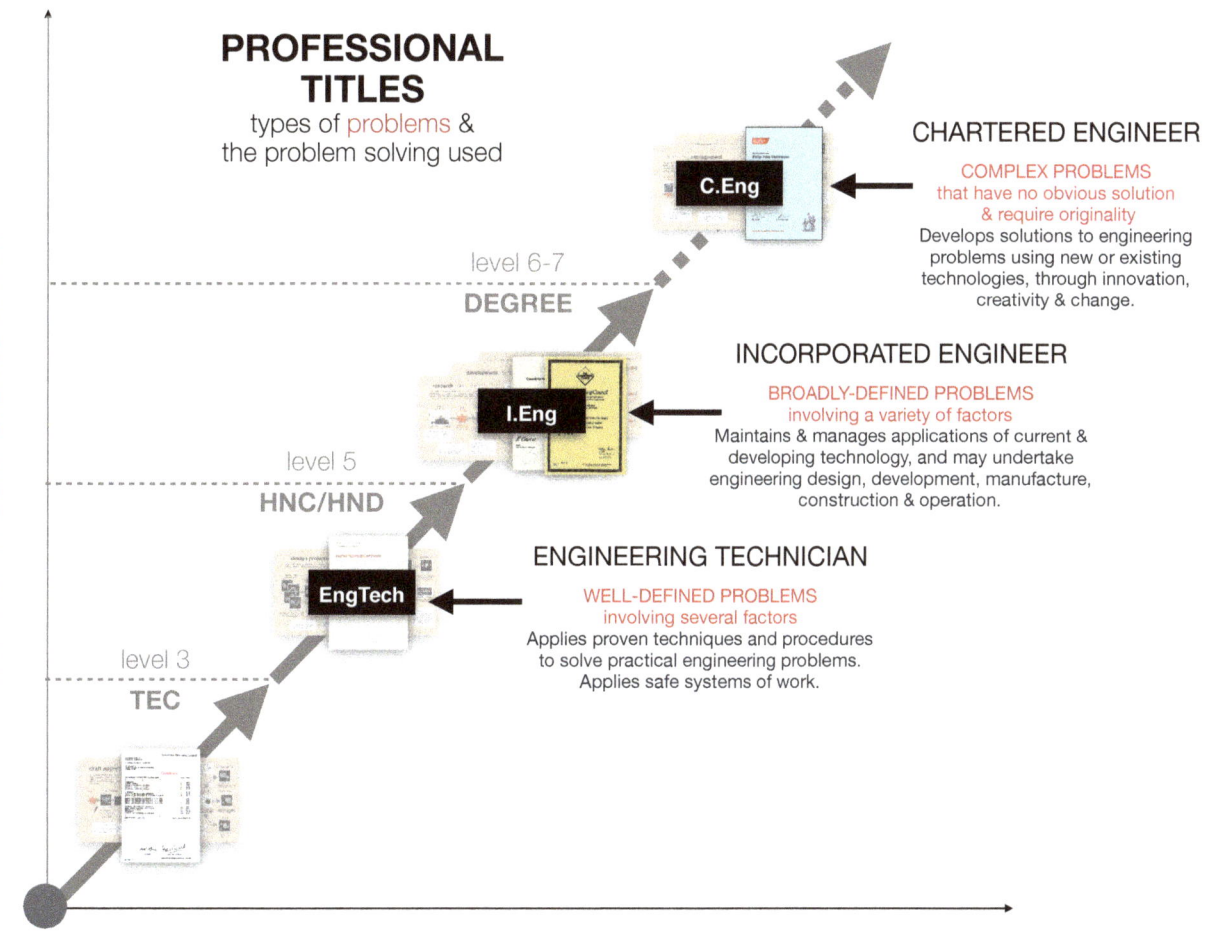

All engineering involves management of some sort, but this increases when we move into leadership positions. The **transition to leadership** will bring a radical shift in responsibilities and capabilities, with a stronger focus on leading and motivating. The engineering manager chapter illustrates how the work or projects change and require us to develop from technical engineers to business managers and leaders of people.

beyond engineering

NEW SKILLS & TOOLS ?

However, adopting an engineering mindset can help you find solutions to business and people challenges. As shown in the engineering manager chapter the tools and drawing can be used to solve **non-engineering problems**. Several of the projects show how innovation can be applied to create new ways of working and develop whole businesses and non engineering products.

 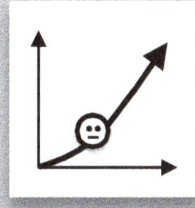

appendix of
book uses

developing more problem solvers

Although this book is a self-learning resource and a handbook for problem solvers, it has also been designed to be a teaching resource. In order to support both students and professionals, much of the book has been based on UK professional standards, engineering qualifications and educational curricula.

Each chapter can be used as a module or lesson or joined together to create a programme or course. The next six pages explain how specific chapters can be used to support primary/secondary education (design & technology), further/higher education (engineering qualifications), and professional development (registration & CPD).

the book as a **learning & teaching resource**

For young people, the book makes the subject of engineering (and design) engaging, and accessible and hopefully brings the subject alive. Helping students see that a career in engineering is exciting, rewarding, and creative. As well as equip them with knowledge, understanding, skills, and process for designing and making.

Engineering education ideally reflects real engineering work and challenges if it intends to prepare students for the realities of the workplace. To meet the demands of employment the content is all linked to a range of real jobs and shows how a set of tools can be used to solve real problems or deliver projects.

The book supports engineers by showing what professional competencies look like in action and when engineers at different levels solve different types of problems. It also gives examples of how engineers can demonstrate these competencies and record their continuous professional development (CPD).

supporting education & the profession

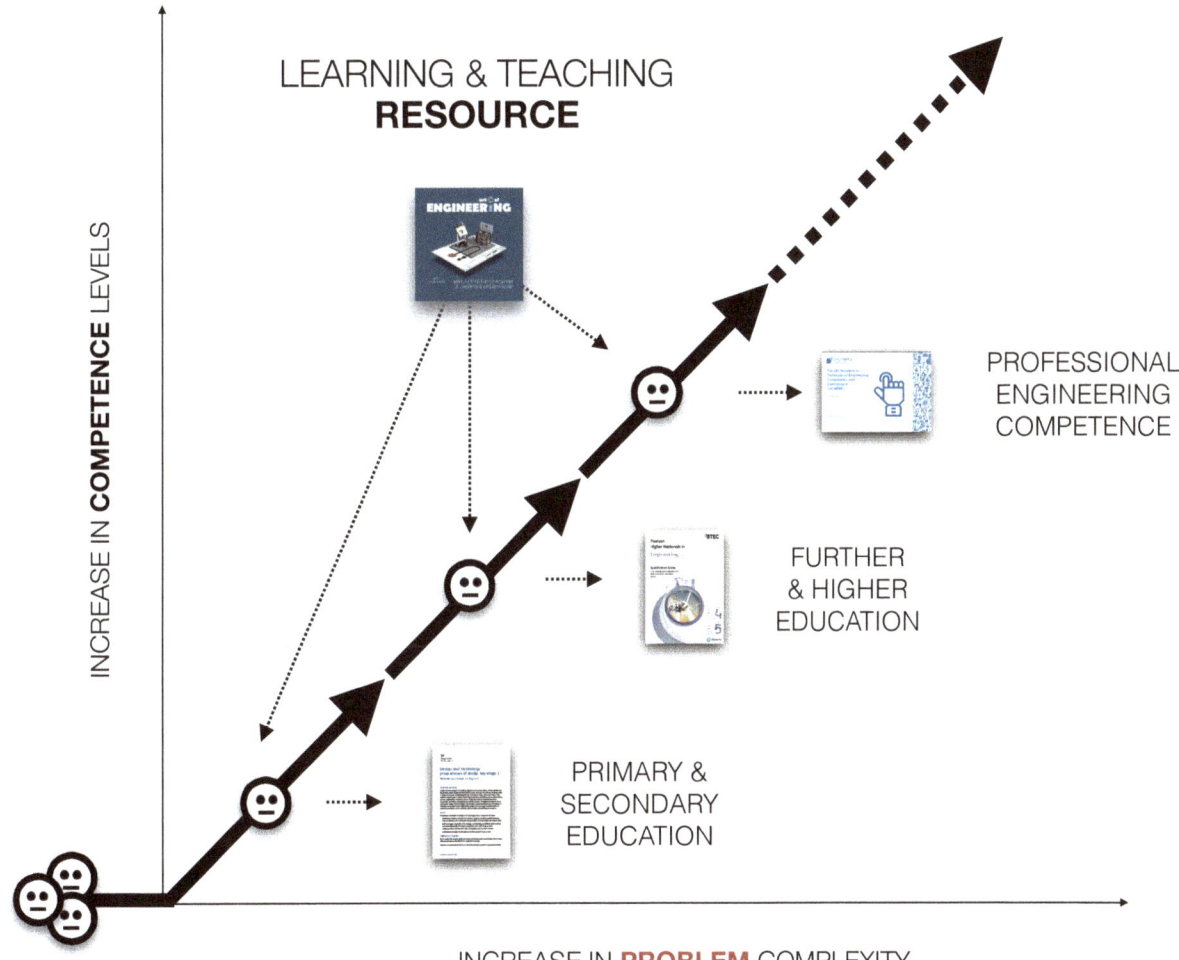

primary & secondary education

The National Curriculum for Design and Technology (D&T) aims to ensure that all pupils can use their creativity and imagination to design and make products that solve real and relevant problems. There is no engineering content in the current curriculum, but the Art of Engineering could help create a link.

The 'mind of an engineer' model links to the curriculum. The nine tools support the development of creative, technical, and practical expertise and help students turn ideas into reality. The use of the tools by professional engineers and their solutions to industry problems can be seen in real examples. Art of Engineering supports the curriculum in the following ways :

knowledge, understanding, skills & process
for designing & making

The introduction to real workplace tools helps the individual engage and experience the realities of everyday problem solving (& engineering) tasks. Hopefully, the book's part on engineers in action will bring this topic back to life and show how exciting a career in engineering can be. The examples of problems being solved and real projects show the iterative process of designing and making.

The new approach to learning : STEAM has been widely discussed in education. It stands for science, technology, engineering, arts, and mathematics, its programs aim to teach students innovation, to think critically, and to use engineering or technology to solve real-world problems. The 'art of engineering' supports the principles of STEAM and could support the future needs of education.

design & technology curriculum

INSPIRING YOUNG PEOPLE
& BRINGING ENGINEER ALIVE

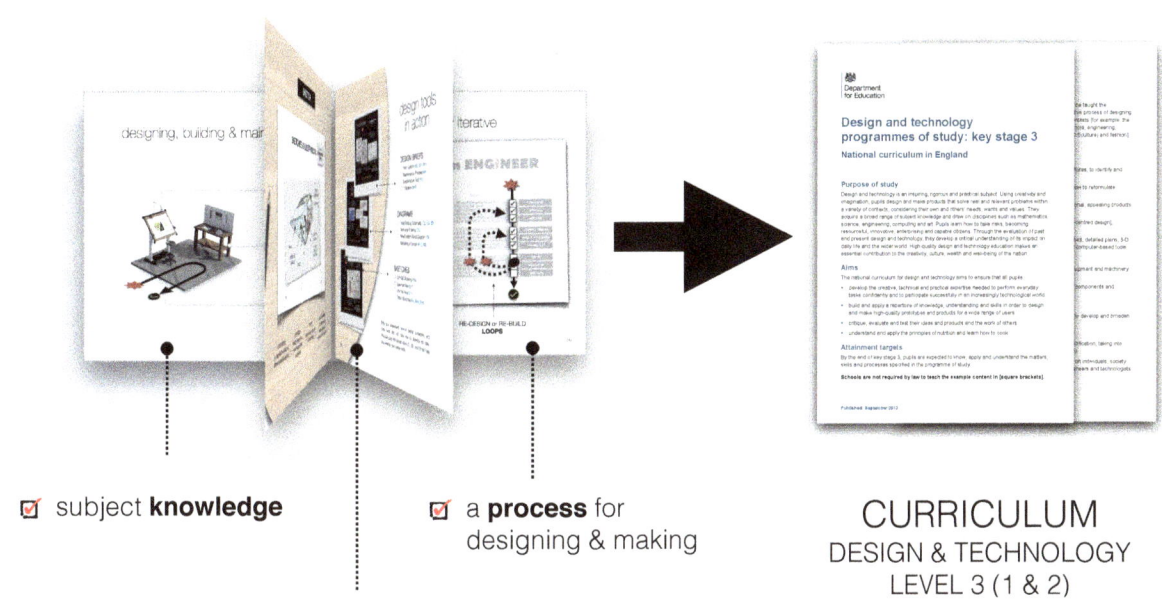

☑ subject **knowledge**

☑ a **process** for designing & making

☑ tools, **skills** & real examples

CURRICULUM
DESIGN & TECHNOLOGY
LEVEL 3 (1 & 2)

further & higher education

Qualifications are designed to reflect the demands of the modern and evolving engineering work, they enable students to apply themselves and give them the skills to succeed in their chosen path. Further and higher education enables students to study the principles and applications of engineering.

The qualifications are aligned to normally included topics/units on fundamental mechanical, electrical/electronic, and mathematical principles, as well as engineering materials, engineering processes, and emerging technologies. The book will help students develop important skills, give them an understanding of engineering practice and career progression

> engineering **design & project** skills & tools
> insights into **jobs**, careers & the workplace

Design and projects are core elements of the book. The design/build/maintain chapters take the student through the process of creating a solution and the engineering project chapter covers the management of an individual design idea to develop a practical solution. Undertaking group projects encourages the development of collaborative and interpersonal skills, the social skills needed for the workplace.

Qualifications provide progression in the workplace, but learning and development are continuous, and lifelong. The jobs and development paths shown in part [ii] give students an insight into the realities of employment, careers, and how engineers develop. The personal notebooks (portfolio pages) also offer a way for students to showcase their talent and to enter, or progress in, employment.

higher national qualifications
& degrees

WORK READY SKILLS
& WORKPLACE PROGRESSION

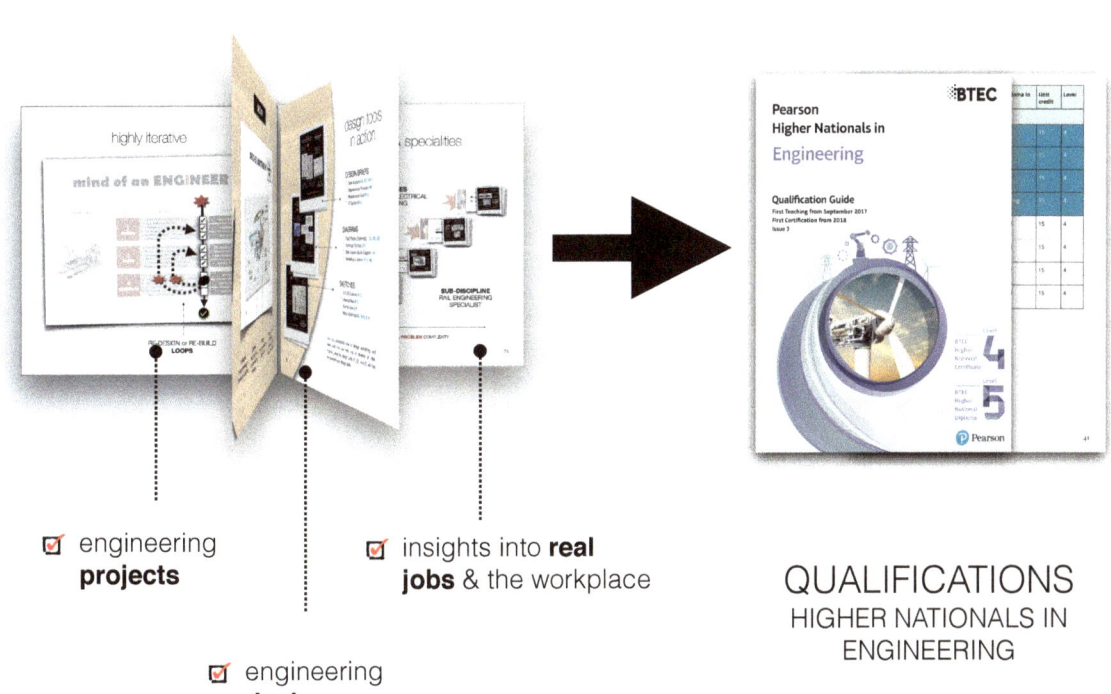

- ☑ engineering **projects**
- ☑ insights into **real jobs** & the workplace
- ☑ engineering **design**

QUALIFICATIONS
HIGHER NATIONALS IN ENGINEERING

professional bodies

Both in the UK and overseas, professional registration gives employers, government, and society confidence in the engineering industry. The standard used to assess the competence and commitment of engineers covers the breadth of the industry and the many different disciplines and specialisms.

Competence is defined as a professional's ability to carry out engineering tasks successfully and safely within their field of practice. Competence needs to be demonstrated in five broad areas : (A) knowledge and understanding, (B) design, development and solving engineering problems, (C) responsibility, management and leadership, (D) communication and interpersonal skills and (E) professional commitment. Art of Engineering shows what these competencies look like in action and can help professionals in the following ways :

> demonstrate **professional competencies**
> & continuous professional **development**

In order to meet the standards, engineers must be able to demonstrate their competences. They must submit evidence of work-acquired experience and underpinning knowledge so that their competence is then assessed. The problem-solving 'polaroid' pages of the book demonstrate competencies in action and show one way of providing evidence for assessment.

Professional registration verifies that an individual can meet the engineering and technological needs of today, while also anticipating the needs of, and impact on, future generations. Continuous professional development(CPD) is essential for maintaining competence and is a key part of being professional. The development paths in the book are examples of development plans, activities, and continuous learning, and CPD in action.

standard for professional engineering
competence & commitment

CONTINUOUS PROFESSIONAL DEVELOPMENT
& DEMONSTRATING COMPETENCE

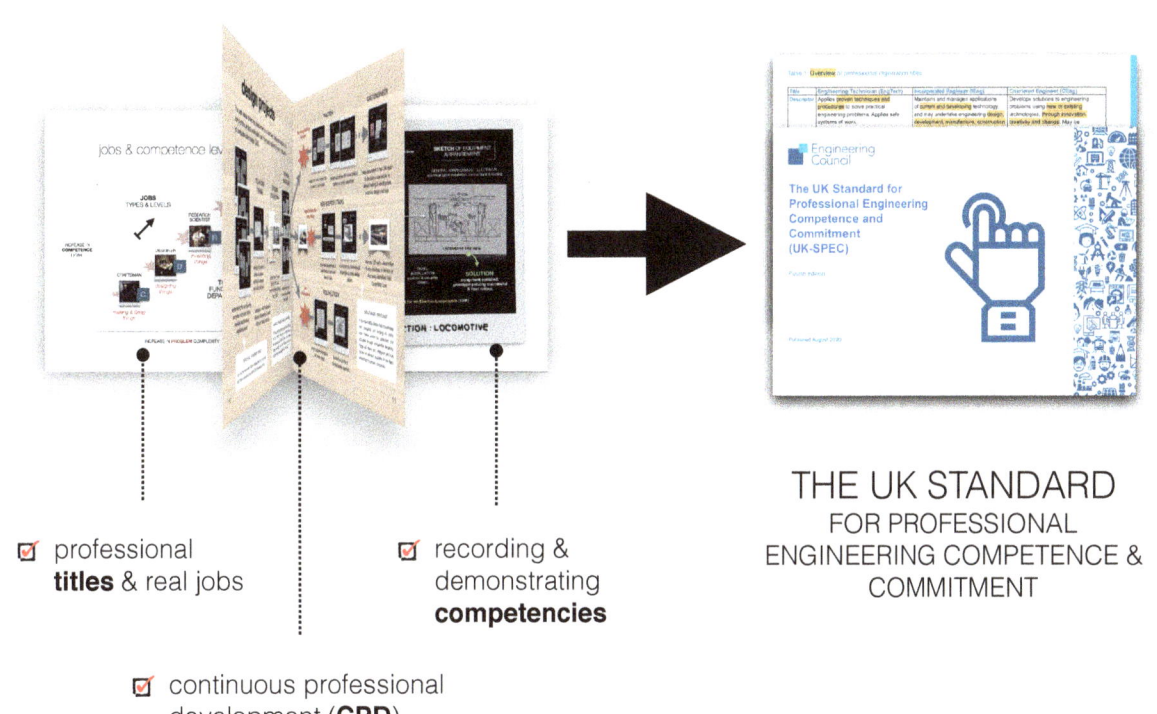

- ☑ professional **titles** & real jobs
- ☑ continuous professional development (**CPD**)
- ☑ recording & demonstrating **competencies**

THE UK STANDARD
FOR PROFESSIONAL ENGINEERING COMPETENCE & COMMITMENT

Qualification and professional recognition are good but they don't get us jobs. Employers are looking to know **who you are**, tangible proof of achievements, and looking to find out what differentiates you from other job candidates. You will need a way to showcase your journey, achievements, and a broad range of skills. If we are going for a job that we haven't done before we need to show that we have the potential to do it and have an ability to learn quickly.

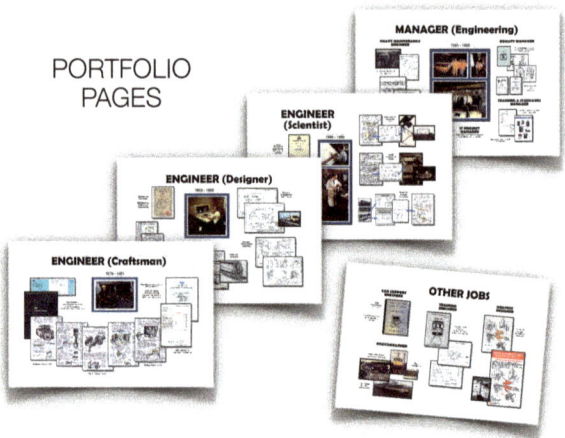

PORTFOLIO PAGES

employment or progression

Personal portfolios (or logbooks) offer a way to visually represent your skills in a creative way and can set you apart from other candidates. A **portfolio** allows you to present concrete evidence of your skills and support claims in your resume. It is a helpful tool during interviews, it helps employers visualise what you could contribute to their team and see your future potential. Creating a portfolio also helps us make sense of our professional and personal journey, and our progress and build self-confidence and resilience.

art of

tools to unleash potential

In 2002, I realised that if I wrote a book, the people I helped would be able to help themselves. So I gave up half my job to become an author, not knowing it would take 13 years to create the book because I wanted to change the definition of what books are !

I had read thousands of **traditional books**, but they were full of thousands of words and didn't always help me, my vision was to give the reader all they needed to learn with one picture and in one book. I would create tools to unleash potential that became **art of** products.

The first book to be developed was 'Art of Enterprise' it would be the most complex subject for me to tackle and the broadest audience, from boardroom to classroom.

But the challenges we would have to overcome and the innovative solutions we went on to develop, would lead to an **interactive book**, a companion mobile **app**, **print books** and a wide variety of applications in multiple sectors. Art of Engineering is the fifth book...

for more information go to
www.artofltd.co.uk

index

index

a
advanced passenger train 112
app development 197
apprenticeship 88-89, 91

b
braking systems 91, 121
book design 196
brochure 57
build 39-41
bus company 194-195
business knowledge 174
business, railway 175

c
career progression 204-205, 220
change management 178-179
chartered engineer 209
collaboration 62
commercialisation 147
communication, internal 192
competence 218
component use 44, 117
computer aided design (CAD) 129, 131
continuous development (CPD) 204-205
craftsman, job 81-83
cross section 43
current collection 139

d
degree in engineering 206-207
design 29-31
design brief 33
design projects 112-113
design office 108-109
design & technology 214-215

designer, job 105-107
development of engineers 201-210
diagram 35
diesel engines 91, 96-97, 99
disciplines 77
documenting 66-67
drawing 3D 127, 156
drawing board 23
drawing office training 118

e
education 212-215
electrical maintenance 92
electrical systems 98, 120
electrification projects 140-145
electrification team 137
employee involvement 188
engineering, what is it 22

f
fault finding 97-98

g
general arrangement 122
group work 62

h
HNC/HND education 205, 214
how to sell it 56
how to use & maintain it 54
how well does it work 52
how will it work 34
how will it be made 44

i
idea communication 36, 66

idea development 36, 64
illustration, technical 123-125
impact on society 56
implementation 66
improvement 60
Improvement teams 189
improving the world 21
industries 72
information for makers 42
innovation 60
inspection system 148
instruction 55
iteration 165
IT system 185

j
job design 190
job type & level 74-75

k
key questions 30, 40, 50
knowledge 86, 112, 138, 174

l
leadership 62, 74
learning & development 202-203
learning from others 91
locomotive, diesel 87
locomotive, electric 139
logbook 88, 91

m
maintain 49-50, 54
maintenance business 180
maintenance depot 84-85
maintenance, preventative 95

make things 90
manager 169-171
managers office 172-173
manufacturing processes 44, 184
manufacturing workshops 114-116
material selection 42, 159
mathematical modelling 162
mechanical maintenance 93-95
mechanical systems 119
mind of an engineer 27

o
on-job learning 91
organisation design 190
overhead line equipment 139-140
overhead line monitoring 143-144

p
pantograph 139-140
pantograph monitoring 141-142
performance model 181
performance management 182
portfolio 220
problem types 72-73
product testing 157
production drawing 42
production planning 46, 61, 183
production process 45
professional bodies 218
professional standards 208, 218
progression 89
projection 43
project reporting 64-67
project plan 47
projects 59
project team 63

prototyping 52, 126

q
quality improvement teams 189
quality, ensuring 46

r
rail industry 76, 80
recruitment, poster 101
report 53
research & investigation 32, 146
research reports 160
research scientist 133-135
r&d lab 136-137

s
safety 37
schematic 35, 97, 98, 100
sell 56
service design 191
sketch 37
specialists 63
specification 128
stages 25
suspension systems 94
strategy 192-193

t
team 63
teamwork 62
technology, new 155
technical drawing 43
technical knowledge 34, 78
technology 34
test, evaluate & improve 52, 151
third rail electrification 145
time & resources to make 46

timescale 47
title block 43
train cleaning 186-187
train coach 111
train operating company 172-173
training material 100, 188

u

understand the problem 32
user requirement 33

w

what does it look like 36
work teams 187
work, engineering 78-79
workplaces 76-77
workshop 23

www.ingramcontent.com/pod-product-compliance
Lightning Source LLC
Chambersburg PA
CBHW042356280426
43661CB00096B/1139